To: Nick

May numbers
cease to aw

BEYOND INFINITY

BEYOND INFINITY

A MatheMATTical Adventure

Charles Ames Fischer

Beyond Infinity: A MatheMATTical Adventure
by Charles Ames Fischer

© 2013 Charles Ames Fischer

All rights reserved.

This is a work of fiction. Names, characters, places, and incidents either are the product of the author's imagination or are used fictitiously. Any resemblance to actual persons, living or dead, events, or locales is entirely coincidental.

No part of this book may be reproduced in any written, electronic, recording, or photocopying without written permission of the publisher or author. The exception would be in the case of brief quotations embodied in the critical articles or reviews and pages where permission is specifically granted by the publisher or author.

Published by

Sienna Books

For more information, visit *BeyondInfinityBook.com*

Cover and Interior Design: Nick Zelinger, *NZGraphics.com*
Consulting: Judith Briles (the Book Shepherd)

ISBN: 978-1-940107-00-4

Library of Congress Control Number: 2013909523

First Edition

Printed in the United States of America

Dedicated to my parents who supported me in everything.

PART ONE

A MatheMATTical Adventure

1

Strange Happenings

Have you noticed anything unusual involving numbers recently? Have you seen your favorite number more than usual? Does there always seem to be an odd number of socks in your drawer? Has something drastic happened on the stock market or to the price of gold? Maybe you keep coming across the numbers 57, 61 or 313. Perhaps it's something simple like receipts that keep adding up to the same number or this book—this page starts on 23.

Well, If you have noticed anything unusual involving numbers, I know why.

It all started normally enough. With just a walk to the bathroom, everything in our entire high school changed. It was September 7th. We had only been in school for a week, and we were still getting used to our new building. Our school was being renovated, so we were currently renting a space downtown. It had previously been occupied by some government agency, one of those that nobody in the town really knows anything about, you know, just a generic "government agency" that all the kids claim is some secretive branch of the CIA or FBI (our town's boring). But that day, September 7th, was when we learned some strange agency really had been there.

I was just going to the bathroom like, you know, sometimes people do. There is a fine art to taking a bathroom break at school: stroll down the hall slowly, stop by the drinking fountain and take three drinks (two of them fake), continue strolling with occasional glimpses at other classrooms and so on, all the way through, well, you know, the rest of the routine. Everything was the same except when I was partway through washing my hands, while I was whistling a Nirvana tune that had caught in my head, I saw a little red light blink from the sink next to me. I thought it was a trick of the light, something reflecting funny, but then it blinked again. I edged over for a closer look. I know that sounds strange, but it was there alongside the sink, one of those little display lights or something that shows that your DVD player is on.

I dried my hands and looked closer, but it just looked like a regular sink. I knew I had seen something, and my curiosity captures me sometimes, really captures me. One time, I suddenly found myself in the mathematics lab after school, talking about quadratic equations because I had heard one of the girls ... wait, that's another story. Anyway, dang curiosity.

So, there I was staring at a sink. Not just staring, crouching underneath, checking the pipes, turning the faucets on and off, and looking down the drain. I'm glad no one came in because I probably looked stupid. I walked around the bathroom and checked for reflections—nothing. I went back to the sink and ran my fingers along the surfaces and felt a hairline ridge on the porcelain. I focused on the side of the sink and there, right there, was a small hole just big enough to jam in a paper clip. I changed angles, and the more I looked in the dim light, the more I could see something mechanical inside. Cool! And no, it wasn't my imagination.

I didn't happen to have a paper clip, but luckily our next class was coming up. I studied the sink a bit more, and couldn't find anything else out of the ordinary. I strolled back to class, having wasted exactly enough time to get back for the homework assignment before class ended. Then I was off to the office where I got a box of paper clips for Mr. Hensen's class—he probably needed them anyway.

I was squirmy and fidgety next class, waiting for an appropriate time to ask to go to the bathroom. If you ask too early, the answer is always 'You should have gone in between classes.' Ask too late, and teachers say to go at the next break. So the trick is to make a chart or graph for each teacher. Whenever someone asks to go the bathroom, mark it down. By plotting the results you can find "the window" as I call it. We were in Mr. Hensen's science class and his window is approximately 11.5 minutes, meaning wait that long after class starts, or before it ends, and he'll let you go. I waited twelve minutes just to be safe and, sure enough, I was out the door.

You might be wondering what I had in mind. That's a good thought, because other than jamming a paper clip in there, I had no plan. I bolted for the bathroom—I hope no one saw me because I probably looked stupid; I take my strolling seriously. There's an art to it, but that's another story.

When I got there, the coast was clear. If other people had calculated the windows for other teachers, they would be here, too. On several occasions I thought of selling the information, but haven't yet.

I unbent the paper clip, poked it in and ... just a little suspense ... and, a small drawer with three buttons slid open, smoothly and without noise. There was no disk or anything inside, but it did expose the interior. The red light I had seen was inside; it must have caught my eye through the hole.

I think there is a law of curiosity about kids and buttons, especially in a situation like this, because before I knew it, before I had even thought about whether it was a good idea or not, I had pushed all three. Bing, bing, bing—just like that. The first two didn't seem to do anything, but when I pressed the third, the light changed from red to green. The dim light in the bathroom changed slightly, and I looked up. That's when the coolest thing to happen in our school really started.

The mirror above the sink was now a computer screen. It still reflected a bit like a mirror, but projected on to the surface was a screen, and under that a complete keyboard. I could see my face in the reflection with letters and numbers across it. There were three empty boxes on the screen with a cursor blinking in the first. That curiosity law kicked in again and I pushed my fingers against the mirror where the buttons for my initials M-J-F were. They appeared in the boxes and then a message ACCESS DENIED flashed for a few seconds and the boxes were empty again.

I heard someone coming, so I pushed the drawer closed and hoped it would erase any trace of the strange computer. Just as the bathroom door opened and some underclassman kid entered, the projected screen and keyboard winked out, and everything in the bathroom went back to normal.

Art class was next, and although we were discussing cool stuff, the computer in the bathroom distracted me. I had to share what happened with someone, and I couldn't wait until lunch, so I resorted to passing notes during class, which is not something I normally do. For starters, I generally find that paying attention in class affects my grades the most. If I pay attention and ask questions, I do better. It's that simple. I know for some people doing all the homework works for them, but I think I'm an auditory learner.

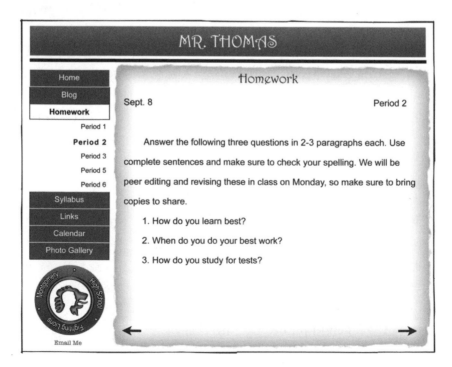

MR. THOMAS

Home
Blog
Homework
Period 1
Period 2
Period 3
Period 5
Period 6
Syllabus
Links
Calendar
Photo Gallery

Email Me

Homework

Sept. 8   Period 2

Answer the following three questions in 2-3 paragraphs each. Use complete sentences and make sure to check your spelling. We will be peer editing and revising these in class on Monday, so make sure to bring copies to share.

1. How do you learn best?

2. When do you do your best work?

3. How do you study for tests?

In tenth grade I did a little experiment. For three weeks I did all of my homework, but hardly paid attention in class. Then I went three weeks where I did no homework and paid complete attention in class. After that, I went three weeks where I didn't pay attention and didn't do my homework, but then studied like mad for tests and quizzes. The results were very clear. Doing all of my homework without the attention in class netted me an 84.7 average over those three weeks. Paying full attention in class netted me a 92.1 average. Just studying or cramming for tests was the worst at an 80.2 average. So, class time for me is extremely important.

Now I do all three, but I understand that listening effectively during class works best for me. Of course, staying organized and turning work in on time is also really important, but that's another story and another self-study that I did. By the way, you should find out what works best for you.

My friend John and I developed a quick system to encode the few notes that we do send. We started with a simple relationship between letters and numbers, assigning the letter "A" the number "1", the letter "B" number "2" and so on. We separated each letter (or number) with a dot and each word with a dash. So a quick "Hi" would be 8•9. Or "Are you going to chess club?" would be written as: 1•18•5—25•15•21—7•15•9•14•7—20•15—3•8•5•19•19—3•12•21•2.

John occasionally got paranoid and would write false messages with false codes, partly for the sake of humor and partly to throw people off. In his craziest moment he borrowed from an episode of *The Simpsons* and wrote a message that when decoded read: "If you're reading this you need a life." He purposely got caught in Mr. Washburn's class and every once in a while we think about the possibility of it being posted in the teacher's lounge and laugh at the idea of one of the teachers bothering to break the code. If I could only get a pic of the reaction!

Piece of paper: $0.0245.

Ink to write obnoxious code: $0.0061.

Picture of teacher at moment of humorous realization: priceless.

Anyway, I knew I needed to send John a message. I wrote out the code: "There's something weird in the bathroom."

John coded back, "Dude, I'm sure I don't want to know." Only then did I realize what I had written. I could see how he might take that the wrong way.

"No. I mean cool weird. Meet me after school at the side hall bathrooms," I wrote back. After reading it, he looked at me and raised one eyebrow. I nodded back to him.

"Just do it," I coded back with the Nike symbol.

2

MatheMATTics

I love math. It's no secret, really. The kids at school call me Mathe-MATT. Some of the mean kids dubbed me Algebrat as well, but luckily, it didn't stick. I don't mind. They just don't understand why I like math so much. I'm not even sure myself—I just constantly see the world in terms of numbers. Even in English class, I manage to think about math. Here's a math poem I wrote in ninth grade.

Holiday Numbers

After 2 whole helpings, 4 desserts and
3 retellings of the Thanksgiving disaster
from 6 years ago, we begged Mom
to take us shopping in the morning.

She said No, but we have her composite number:
12 Pleases and she caves in,
defeated. Dad's an 8, Grandpa's a 6;
My sister Claire is irrational.

Early the next morning, the 5 of us,
our stomachs bulging like trapezoids,
piled into the RAV 4 and drove down
route 99 right to the Square 1 Mall.

The associative property of the mall on
Black Friday is amazing: mixed numbers
of people in huge lines are already
angling around even before 5 AM.

With the sales, every product becomes a factor,
even the imaginary: Claire tried
to fit into size 5 jeans. I guess 70% off is
appealing when your pants are too tight.

After buying a lot of stuff we didn't need,
we carried 8 stuffed bags to our car
parked in section 22, drove 6 blocks and
stopped at the 7-11 for gas.

When we arrived back at our house and
stumbled through the door, we were
greeted by our 3 dogs. Exhausted,
we managed 1 "Thanks, Mom."

Math is beautiful and simple. It follows rules, has predictable order and almost always has a specific correct answer. There's no speculating what characters are thinking in English class, no interpreting what primary resources mean in social studies, no hypothesizing about what will happen with science experiments. With math there *is* a correct answer. I know in quadratic equations and higher functions there are multiple right answers, but I mean in general.

Math is a hobby for me. I don't just mean I sit around doing math problems out of a textbook, although I do that sometimes. When I get bored, I turn to math. For example, in Ms. Briggs' class last year, I zoned out sometimes, so I began to count how many times she said "um" or "ah." Her record is 211 in one class; that's an average of 4.79 per minute. I thought she was going to be the all-time record holder until Mr. Umberger subbed one day. In our 44-minute period he had 122 "ums" and 209 "ahs" for a total of 331 (almost my favorite number). That's 7.52 per minute or roughly one every 7.98 seconds. He currently holds the record and has been awarded the prize of *Most Able to Talk a Lot and Not Say Anything*.

Inside the domain of mathematics there are a lot of really interesting things to ponder. Numbers, themselves, for example, may seem boring, but they can be fascinating on many levels. For starters, there are different ways of approaching them. Perhaps the most obvious is that numbers can count quantities. People a long time ago realized that three pineapples, three llamas, three hiccups and three Neanderthals have something in common—that is, their three ness When numbers are used to count quantities this way, they are called cardinal numbers.

Numbers can also represent an order of events. When we say *first* or *second*, or *fifteenth*, we are using numbers to create an order instead of a quantity. For example, when we say *first*, we are talking about the fastest or best. But *first* does not tell you the quantities of things. You could be the first out of ten people, twenty, a thousand, or a million people. *First* tells you that you are ranked the highest in the group, but it does not indicate how large the group is. When we use numbers to create order, we call them ordinal numbers.

Ranking things in ordinal order at first might not seem all that important or interesting, but we use them all the time for things like anniversaries, birthdays, and sporting events. Think about coming in first place in a chess tournament, third place in a spelling bee, or even winning silver in the Olympic games. For a simple example, take an oak tree, which is always *first* an acorn, then *second* a sapling and then *third* a tree. Or a butterfly is always *first* an egg, then a larva, then a pupa and then an adult. The order doesn't change, *can't* change.

Then, the question arises: Which came first the chicken or the egg? It's pretty easy, actually. The egg obviously evolved before chickens. The egg as a biological phenomenon was around a long time before chickens evolved. What I think most people really mean to ask is: Which came first, the chicken or the chicken egg? Sorry, I can't help you with that one.

Matt,

Your grandma is back from England and you promised to take her to bingo this week.

Love,
Mom

Numbers have more personality than most people imagine. In British bingo, for example, some numbers have names based on their shapes. The number 2 is called One Little Duck, or One Little Swan, because it can look like a duck or swan on a pond. Personally, I think two looks like a cobra. 7 is called One Little Crutch, 8 is One Fat Lady and 11 is known as Skinny Legs. Then there are combinations like 87: Fat Lady With A Crutch, 27: Duck With A Crutch, or 82: Fat Lady With A Duck. By extending this idea, you could get 727: Little Duck with Crutches, or 222: Cobra Attacking Two Ducks, or even something more complicated like 7872112: Fat Lady With Crutches Watching a Cobra Attack a Duck with Skinny Legs.

Some of the Bingo numbers have fun rhymes that make them more interesting, like:

- 4 – Knock on the Door
- 25 – Duck and Dive
- 30 – Your Face is Dirty
- 31 – Get Up and Run
- 42 – Winnie the Pooh
- 46 – Up to Tricks
- 74 – Candy Store

I used to make up stories in math class based on the answers on our worksheets. For example, the first few answers might be 2, 49, 30, 77, 478 and 46. I would imagine a duck character who just woke up with his face all dirty. He went waddling down Crutches Avenue to find house number 478 to play some tricks on whoever lived there. Not the best plot in the world, but it made writing in second grade easier, although my teacher got sick of duck stories.

2
Two as a swan

2
Two as a swan in fog

2
Two as a cobra

Thinking about numbers in very basic counting terms brings to mind that numbers create shapes. I'm not referring to how they are written, like 3 or 5 or 7, although that's interesting too, since they have changed over time. Instead, I'm referring to how they can be arranged or what formations they can make. The Greeks, for example, called them figurate numbers, or figured numbers. If you represent each number as a stone and then arrange those stones into various shapes, you can discover certain properties of various numbers. This might be where we got the phrase "figuring something out."

See that the number one can only be arranged in one way as a single stone:

O

Two can be counted or figured in more than one form (such as placing one stone on top of the other), but the basic form would be to place them next to each other to form a pair:

OO

By moving the second stone around, we can demonstrate a lot of mathematical concepts, such as a*ddition, difference, opposition,* and *length:*

O --------- length --------- O

The number three has even more choices. They could still be stacked on top of each other, of course, or they could be placed in a line segment, which can create a few more new ideas, such as *fracture* or *fraction, midpoint* or *median,* and *equal* and *unequal.*

O   O   O

But the powerful difference between arranging two stones or three stones is moving them *out of line.* When this happens, we can see that three stones can create different angles, or tri-angles. Interestingly, there are three types of triangles, depending on how we move the stones around. We can create triangles with all equal sides, *equilateral triangles*; those with two equal sides, *isosceles triangles*; and those with no equal sides, known as *scalene triangles.* Arranging three stones in these triangles also creates two major new ideas: *perimeter* and *area.* Arranging four stones can create even more shapes, but that's another story …

Anyway, back to the main account. The rest of the day I was wondering about the sink and what had happened. What was it doing there? Why was it there in the first place? What was in this building before our school district rented the space? What was the access code for?

By the end of the day, I had daydreamed the strange access panel into a million different things. Okay, not a million, but a lot. In my most imaginative moments, I thought of the panel as belonging to one of the nuclear missile silos left over from the Cold War, or a biohazard analysis device for scientists working with biochemical weaponry, or even a special agent check-in terminal for someone like James Bond or Jason Bourne. My personal favorite was that it was a multi-million dollar soap dispenser. Ah, tax dollars at work, right? On my more practical daydreams, I thought it might be a water quality control panel or some kind of water pressure gauge or thermostat. But none of them made sense. In fact, *nothing* about it really made sense. After all, a computer panel in a sink? Water and computers don't go well together.

All day I thought about how could I crack that access code. Was there a clue about what three numbers or letters to type in? Could I access the CPU and run a software program? Could John and I pry open the sink without damaging it? Maybe then we could take the computer apart and either reprogram it, download the information, or perhaps install new hardware.

John met me after school at the bathroom with a scowl on his face. "Dude," (he says dude a lot), "this cool thing better be ninja." And he's not always known for his eloquence. He swiped his fingers through his brown hair. Sometimes he bleaches his hair, but on this occasion, it was his normal brown. I think he secretly wants to be a surfer, which explains why he says dude a lot. But he's not very coordinated. He tried skateboarding a few times and nearly died on the sidewalk outside his house in a disastrous incident with a garden hose.

I held up my hand, "Hold on." I went into the bathroom quickly and checked to make sure no one was in there. It was clear. I came back out. "Do you have a paper clip?"

"What? Dude, I'm missing a ride home for this. What's so cool?"

I took that as a no. I fished through my pockets and luckily still had the paper clip I had used earlier. "Come on," I said checking the hall for anyone approaching, "I'll show you."

I went directly over to the second sink. "Check this out." I poked the paper clip in the small hole and once again the small three buttons opened up. I pressed the third button and watched John in the mirror as the screen appeared and covered his face with the three boxes.

"What the …" John trailed off. He looked around the bathroom, his eyes tracing the edges of the walls. "What is that?"

"Some kind of computer," I muttered.

"What do the other two buttons do?" he asked.

"I'm not sure yet, but this thing must be something ninja." I typed in his initials on the mirror screen and ACCESS DENIED appeared again, just as it had for me. "The question for me is, what do we get access to?"

"Dude, this could be the school computer! I could change all my 'D's to something like 'B's!" John's a smart kid, but he doesn't try very hard in school.

"John, man, why would the school computer be in the boys' bathroom? Miss Chavez does our report cards. You think she sneaks in here to take attendance every day?"

"Well, what is it, then?"

I shrugged my shoulders. I wish I could have given him a decent answer, but I was stumped too. Why is a computer built into a sink in the bathroom? I looked over the whole sink again and came back to the small panel with the three buttons. "What do you think these other two buttons are for?"

John bent down to look closer. "I'm not sure." He looked back up at the mirror screen. "What passwords have you tried?"

"Just a few," I said, "I discovered it this afternoon and only had a little time. It's weird, huh? I keep wondering what it's doing in here."

"Can we break into this thing?" John said as he examined the panel and the mirror. "There's no input jack. No USB, no A/V, nothing. We couldn't run a diagnostic program or anything unless we wired it directly into the …" His head tilted. "That's weird. Look at this." He was down on his knees pointing in through the small gap in the sink left by the panel. Inside was a circuit board, but that's not where he was pointing. Part of the faucet pipe had been replaced directly above the circuitry. It looked like it was made of glass or clear plastic. "That's pretty ninja, man. I was thinking of just breaking this sink and taking the drive and parts out. Then we could just hook them up to my computer and figure things out. But look, if we try to force anything, I'll bet that piece will break and water will pour out over everything."

John was right, at least it looked as though he were right. It was pretty ingenious, actually. No input devices. No breaking in for the hard drive. Either we would have to crack the password, or we would have to shut the water down, drain the pipes and then break it open. Besides potentially getting into serious trouble at school, something told me that shutting the water down would trigger an alarm or a shut down code or something. Maybe that's what those other buttons indicated.

"There's no way we can gain access to this thing then," he said standing up. He ran his fingers through his hair. "Unless you want to sit here all day and …" We heard a noise out in the hall. I instantly closed the panel and the mirror screen faded out, just as Derek Morton came in. We must have looked a bit awkward standing there because he greeted us with, "What the hell are you girls looking at?" This is about as nice as Derek gets, at least to me.

I nudged John because he looked like he was about to say something stupid, and Derek has punched him on several occasions in the past few years. When we were out in the hall, I smiled. "I have an idea, John, but we need a few other people."

3

One, Two, Three, Six

My old math books have typically stopped at cardinal and ordinal numbers, but I have since realized that there is another level beyond thinking about numbers simply as quantities or order. This deeper level is thinking about numbers as a process that combines both the ideas of quantity and order.

The counting number one comes first and it represents a single quantity. Two obviously comes second and represents the quantity of two, but here's the important part: two *follows* one because one cannot exist by itself in isolation. Imagine if we could ONLY count to one because that was the only number we had. We couldn't have two hands, two ears, two eyes, or two of anything for that matter. Try to imagine darkness without light, up without down, or fast without slow. In these cases, as soon as you have something that is fast, there must be something to compare it to. Something must be slower in order to "make" it fast or faster. Something must be darker in order for something else to be light or lighter. In order for there to be an "up," there must be something else that is "down."

That something else is two.

So when something comes into being, its opposite comes as well, because the idea of *left* can't exist without the idea of *right*. Or the idea of *hot* can't exist without the idea of *cool* or *cold* or *warm* or any number of other words that describe a variation of temperature. For example, let's say the world was exactly fifty-five degrees Fahrenheit everywhere, without variation at all. I know that's impossible because of wind, shade, rain, sunlight, and all that, but just imagine. Now imagine that a cloud passes over and the region below gets slightly less sunlight. Now that region drops to fifty-four degrees. Instantly we have a difference, we have a *cooler and* we have a *warmer*. They both have to exist because the terms are relative to each other.

So the number two is second and it's a quantity of two things, but it's also part of a process.

Three comes next and is a little more complicated. I'll start with a language example. When we are asked a basic question, we can answer yes or no. Yes, of course, means acceptance and no means denial. There are the two opposites again. But there are always other choices for things, options or different directions that things can take. In this example, an answer could be maybe, which is not a yes or a no. Notice that maybe would exist in between or in the middle of yes and no.

An example of three as part of a process is to first think about light and dark. When they mix, we get blended grays. You can think about day and night, and then the transition periods known as dawn and dusk. So the number three is the multiple blending of the two opposites that came before it. That is why the numbers are in the order that we know: one, then two, then three. However, since three exists in the space between the two opposites, we might actually consider counting 1, 3, 2.

An example with pictures might help illustrate. Imagine if "hot" were represented by the number 1—then "cold" could be represented by the number 2. When the two extremes interact, there's a whole range of in-betweens, such as warm, moderate, temperate, tepid, cool, chilly, and so on.

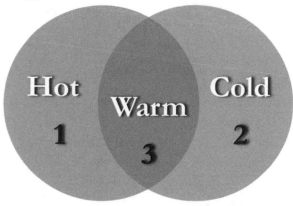

So, three is the middle ground between two extremes. Sometimes it can be seen as the overlapping of the two extremes, as in the example with hot and cold. But sometimes three shows up as the binding between the two opposites. For example, take the opposites of *buyer* and *seller*. What binds them together? In other words, what would cause them to interact? Well, an object or service, whatever is being sold. Or another way of thinking about it is that they would exchange money. Or take the opposites *projector* and *screen*. What third thing unifies them? Well, the movie, of course, would be one answer, since the purpose of a screen and a projector is to show movies. But another answer could be the audience, since a screen and projector are for showing to groups of people.

Take the opposites *plaintiff* and *defendant*. What binds them together? Well, again a first answer, and perhaps the most obvious, is the lawsuit. But another way of looking at it could be that the judge is the binding force. The jury could be a potential answer to what binds a plaintiff and a defendant together, and so could a more generic answer of "the law."

Take any pair of opposites and search for the middle ground between them. You'll see that there are many possibilities depending on how you think the two things are bound together. Think about action, reaction and resultant, or thesis, anti-thesis and synthesis. Two opposites that interact create numerous possibilities, just as a man and a woman can have many children. Three marks the boundary or gateway to the many. So, one is singular, two is dual and three is multiple.

Here are some examples of three from my numbers collection:

1. Positive, negative, neutral
2. Protons, electrons, neutrons
3. Man, woman, child
4. Red, white, blue
5. Past, present, future
6. Late, early, on time
7. Writer, reader, book

8. Friend, foe, neutral

9. The Great Pyramids of: Khufu, Cephron, Menkaure

10. Father, Son, Holy Ghost

11. Acid, alkali, neutral (water)

12. Id, ego, super-ego

13. Win, lose, tie

14. Beginning, middle, end

15. Fat, protein, carbohydrate

16. The Three Musketeers

17. Length, width, height

18. Stoplights: green, yellow, red

19. "On your mark, get set, go!"

20. Three strikes in baseball

21. Olympic medals: gold, silver, bronze

Want to see 1-2-3 in action together? Pick a number, any number. It can be a single digit or a string. Just make sure to include zero as an even number. Count the number of even digits, the number of odd digits, and then the total number of digits. Use the new numbers in the answer and continue the process until it repeats. Guess what happens? 1-2-3 every time. Here are two examples using the digit 6 and the number 547,484 to illustrate:

Digit: 6
Number of evens: 1
Number of odds: 0
Total: 1

New Number: 101
Number of evens: 1
Number of odds: 2
Total: 3

New Number: 123
Number of evens: 1
Number of odds: 2
Total: 3
Sequence: 1-2-3

Number: 547484
Number of evens: 4
Number of odds: 2
Total: 6

New Number: 426
Number of evens: 3
Number of odds: 0
Total: 3

New Number: 303
Number of evens: 1
Number of odds: 2
Total: 3
Sequence: 1-2-3

Not a crazy enough example to convince you? How about a number along the lines of 78,235,647,901,445,933? Even this huge number succumbs to the power of one, two, three.

Number: 78235647901445933
Number of evens: 7
Number of odds: 10
Total: 17

New Number: 71017
Number of evens: 1
Number of odds: 4
Total: 5

New Number: 145
Number of evens: 1
Number of odds: 2
Total: 3

New Number: 123
Number of evens: 1
Number of odds: 2
Total: 3
Sequence: 1-2-3

So, the other day, I was thinking about this 1-2-3 thing and how they add up to the number six, which is a number that will soon become important to this story. Six is such a ninja number, but it often gets a bad rap, probably because of its association with 666, or because of the phrase "six feet under," meaning dead and buried. Or because all insects have six legs and a lot of people hate bugs. Six is the atomic number of carbon, which is a building block of the universe, or in this case, our code-breaking group. Six is the number of points on the Star of David, and day six is when God created Man. A six-sided object is a hexagon and those can be seen in a lot of crystals like the basalt columns at Giant's Causeway in Ireland. I'm still fascinated that pretty much all snowflakes are based on the number six. And I definitely don't want to leave out Six Flags amusement park.

Six is a great number. In fact, it's perfect. I'm serious; it really is a perfect number—as defined by mathematicians. A perfect number is one where the sum of the positive factors (not including itself) equals the number. In this case, six can be divided by 1, 2 and 3. Add those together and the sum is equal to six. Perfect numbers are pretty rare. The next perfect number is 28, and another doesn't occur until 496.

So, in my thinking, I needed to form a group of six people in order to break this code.

In geometry, six identical circles fit precisely around a seventh circle. You can see this by arranging six pennies around one central penny, or six tennis balls around a seventh. Drawing line segments through the centers of the circles can mark the corners or vertices of a hexagon. This six-around-one pattern makes the hexagon perfect for tiling floors because the corners interlock efficiently. Bees use the hexagon's strength and efficiency to make their hives.

By the way, I have a working theory about the number 666. It's usually called the number of the beast from the reference from the Book of Revelation. I think it might be the number of addiction. First, regular dice are six-sided and are some of the oldest forms of gambling in the world. The sum of the numbers 1 to 36 numbers is 666, and 36 reminds me of rolling two dice, since there are six times six, or 36 combinations of the two dice. Second, all the numbers on a roulette table mysteriously add up to 666. Next, beer comes in six-packs and "six-pack abs" refers to people who work out all the time. As far as I'm concerned, they're addicted to exercise.

I've thought about why three sixes as well, and I think it has to do with six on three different levels. They could be id, ego and superego, or maybe the past, present and future. I guess what I'm getting at is that the Beast might be within oneself. Anyway, I said it was a working theory.

To help break the access code, I collected together five of my closest friends. John was one, of course. He's pretty good with computers, and besides, he and I have been friends since kindergarten. I next thought of our friend Kelsey, or Kelsie, as she now spells it. She has a good head on her shoulders and is always looking out for people. She's a risk taker, but only when she has worked out how everything can be safe. I'm glad she decided to join us, because without her this whole thing would have failed.

Jamie and Thomas, who are brothers, were next. Jamie is the older one and is in our senior class. Thomas is a junior. They moved to our town when I was in fourth grade and we've been friends ever since. Jamie is one of those remarkable people who can recreate just about anything he sees. I'm amazed at what he can draw, paint, and sculpt. When I try anything beyond stick figures, which I must say I think I have mastered, then my hands just don't seem to obey. Thomas is the athlete amongst us. He plays soccer in the fall, basketball in the winter and runs track in the spring.

Ari was the last to join. I thought of asking a few other people, but I kept coming back to Ari Goldman. I first met Ari on the playground back in third grade. I was getting harassed by a group of kids who had trapped me in the jungle gym. I don't know how he did it, but he just walked up to the group and said, "Let him out." There was a strength or determination in his voice and the entire gang of kids just walked away. It's not that Ari was bigger than any of them—he was just as scrawny and skinny as I was. Somehow his voice had enough power and they just stopped. We still are scrawny and skinny, by the way.

I got to thinking about the seven circles as a symbol for our situation. The six outer circles represented the six of us, and the seventh circle represented our problem (or solution). After I showed the others, we actually adopted the six-around-one emblem as our symbol. Jamie eventually made t-shirts for all of us as well.

So our group of six was formed. We needed a ninja name for ourselves, but our initial ideas were pretty lame. John suggested the Six Amigos and the Six Musketeers, both of which we shot down immediately. Thomas wanted us to be named after a sports team, mainly for the connection to teamwork, but the rest of us didn't really follow sports enough to be excited by the idea. Ari wanted something bold, but kept drawing from violent video games. He wanted things like The Smashers, or his favorite, The Breakers, which almost won because we were trying to break a code.

Kelsie came up with the idea of calling us simply The Company. She said she got the idea from Sir Arthur Conan Doyle's book *The White Company*. After she described the book to us, we all liked it. I liked it because it was simple. John and Thomas liked the team aspect. Ari liked that the name came from a novel about war, so he was happy. By the way, we didn't call ourselves the *White* Company because we weren't all white.

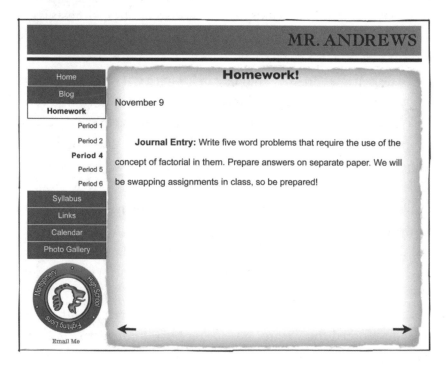

MR. ANDREWS

Home
Blog
Homework
Period 1
Period 2
Period 4
Period 5
Period 6
Syllabus
Links
Calendar
Photo Gallery

Email Me

Homework!

November 9

Journal Entry: Write five word problems that require the use of the concept of factorial in them. Prepare answers on separate paper. We will be swapping assignments in class, so be prepared!

Assuming that the three boxes on the mirror screen in the bathroom were to be filled with either numbers or letters and not symbols, there were basically 36 possibilities per box: 26 letters and 10 digits. That's $36 \times 36 \times 36$, or 46,656 combinations. Subtracting out some improbable combinations such as 111, AAA, or 123, and accounting for multiple attempts during breaks, interruptions, recess, and subtracting for after-school visits, I estimated we needed 12,000 bathroom breaks during school hours. That would mean in the remaining days of school I would have to go to the bathroom 71 times a day. Even I couldn't possibly excuse myself that many times, which is why The Company was formed.

Our group crystallized into action and purpose. With five of us working steadily it would mean 14 trips each to the boys' bathroom every day, and even then there was the distinct possibility that we wouldn't crack the code until very late in the year. So I had to increase the numbers. My goal was to have each of us make 21 trips a day to the bathroom. Twenty-one's a great number, too—mainly because it's formed from three sevens.

You're probably wondering how I was going to make this happen, how we could possibly get excused from all of our classes twice a class, but you have to understand that my mind works really fast sometimes. Before I had even gotten home that day my plan was hatched. Doctors' notes. We would all get doctors' notes, and then the teachers couldn't do much about it even if they wanted to.

Of course, the trick would be to make it realistic. It wouldn't take much for someone to figure out forgeries. Or, for that matter, the strange "coincidence" that all of us suddenly were afflicted by the same thing. And we couldn't get them signed by the same doctor either. No, if this was to work, we would have to convince our parents, the doctors, and the school administrators that this was all legitimate.

So the five of us hit the library (Jamie couldn't make it, so I promised I would explain everything carefully). Research was the key. With the right persuasion, the doctors would sign the notes. There were two things on our side: a trip to the doctor's office is predictable, and no doctor can handle the exotic or strange. As you'll see, those two things worked like a charm. How do charms work, anyway? I wish I knew because I certainly could have used one in a matter of weeks.

After the library, all of us stayed overnight at my house. We filled Jamie in as much as possible, although he can be distracted easily, so I worried that he didn't get everything—which he didn't as I later found out. I'll tell you about it in a minute. I explained my plan to them and they agreed, except for Jamie. He had his own plan.

Here's what happened. The next day we all reported the same thing to our parents and told them that something tasted weird last night, but we just thought it was the fruit because we had nothing to compare it to. We complained of pain in the lower abdominal region—the exact location is extremely important. With descriptions of sharper pains and good acting over the next few days, it wasn't too long before we were brought to the doctor's office.

Remember, there were two things on our side: a trip to the doctor's office is predictable, and no doctor can handle the exotic or strange. With a trip to the office comes the inevitable urine test, which we gladly—but hopefully not too eagerly—agreed to. We explained the pains to the doctors, being careful to get the location exactly right. And that's where Jamie messed up. He described the pains in the wrong place and ended up being referred to a gastroenterologist. They made him take a special medicine that gave him really serious gas. He wasn't very popular at school the next few weeks.

Anyway, the urine test was the first part of the plan. The second started when the doctor got the test results back. Calmly we told him that we had had a taste-testing session where we tried several strange mushrooms, a dozen or so rare fruits, like abui, biriba, and canistel, and other assorted items. The doctors in my office anyway, scurried

around for answers. After a few hours, they told my mom that everything was going to be fine, but they didn't really know what effect those
things would have on my system. They suggested that I drink a lot of
water and try to flush whatever toxins there were out of my system.

Perfect.

4

Spacing Out

One of my favorite pastimes is playing around with numbers. It's kind of like doodling, but instead of drawing, I manipulate numbers. For example, I might try using basic mathematical operations to work with a single digit in order to generate all of the numbers from 1 to 100. I'll give you some examples using the number four:

$$4 \div 4 = 1$$
$$(4 + 4) \div 4 = 2$$
$$4 - (4 \div 4) = 3$$

For larger numbers use something called factorial, which in the case of four is written like this: "4!" Factorial means take a positive number and multiply it by all of the other numbers counting down to one. So 4! means $4 \times 3 \times 2 \times 1$, which equals 24. Another example is 6!, or $6 \times 5 \times 4 \times 3 \times 2 \times 1$, which equals 720. The results get really big very fast. Eight factorial jumps to 40,320 and 15! is an incredible 1,307,674,368,000.

So for larger numbers:

$$4! \times 4 = 96$$
$$4! \times 4 + (4 \div 4) = 97$$
$$4! \times 4 + 4 = 100$$

You can make this pastime extra challenging by also requiring a specific number of digits, in this case, always using four "4"s. Try generating every number from 1 to 100 by using exactly four "4"s. It's not easy. Try other combinations too, like seven "7"s or six "6"s. It might not seem like great fun to some people, but in my mind, it's no different than doing Sudoku or crossword puzzles.

Factorial, by the way, is used in combinations and permutations, and shows up a lot in probability questions. Essentially, when things can be arranged or rearranged, they have a chance (sorry about the pun) of using factorial. For example, if you were asked: How many different ways can the numbers 1, 2, 3 be arranged? The answer can be found using factorial, in this case 3!, or $3 \times 2 \times 1 = 6$. Here's why. At first you can choose one of the three numbers to start, which leaves two left. Once you pick the second number, only one remains. That's $3 \times 2 \times 1$. Three choices, then two choices, then one choice.

I know such a pastime may seem pretty nerdy to some people, but everyone has their distractions, like movies, books, painful memories, or whatever. For some, the distraction is sports or more likely watching sports. That's my dad for you. He is the poster child (poster adult?) for couch potatoes everywhere. Dad loves watching sports of any kind, particularly hockey and football. He played football and baseball in college and kept playing in various men's leagues until he blew out his knee.

Dad tried to get me into sports when I was younger, so I was forced to try hockey, baseball, football, soccer, tennis and even lacrosse. I had remarkably embarrassing moments in most of them. One time playing football, I got the ball and just blindly ran until I noticed people on the sidelines yelling at me to run the other way. I changed direction and ran that way for a while until there was more yelling, so I turned left and ran some more … right off the side of the field, past some people in folding chairs and into the parking lot. I only stopped because a tiny kid riding a tricycle knocked me over. Embarrassing. I fumbled the ball too. That was the end of my football career. And I don't want to say anything about my attempt at playing hockey, except to say that as far as ice and I are concerned, I should stick to ice cubes.

I finally settled on soccer because the game made more sense to me for some reason and I can run pretty well. I was never great at the skills part of the game, but I could play defense well because I was good at "finding the angles" as Coach called it. I played on our travel team until high school when I was the last person cut from the JV team. I haven't played much since then.

My mom's distraction is chess. She plays almost constantly with some online friends. She even has a computer set up in the kitchen so that she can walk by it, study the board and make a move even when she's puttering around the house. She has a collection of chess books that I rummage through sometimes. I'm always amazed that there could be so much to write about, but there they are entire books dedicated to single famous games of chess played by people like Bobby Fischer, Anatoly Karpov, Susan Polgar, and Garry Kasparov.

I have played a lot of chess over the years. Mom got me started when I was four. I was a local chess champ at eight and played in numerous tournaments until I was twelve. Then the stress got to me, and I didn't like the game anymore. Too much competition and not enough fun or love of the game. I stopped playing almost completely until high school when I joined the chess club just for fun. I enjoy the game again, but now just as something fun to do.

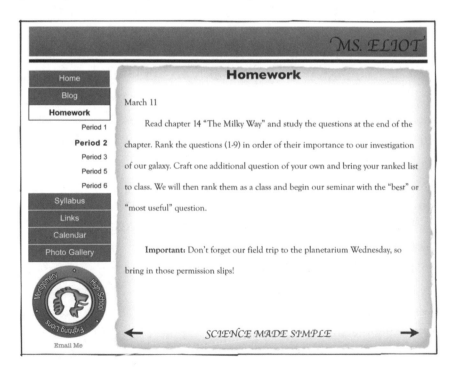

MS. ELIOT

Home
Blog
Homework
 Period 1
Period 2
 Period 3
 Period 5
 Period 6
Syllabus
Links
Calendar
Photo Gallery

Email Me

Homework

March 11

Read chapter 14 "The Milky Way" and study the questions at the end of the chapter. Rank the questions (1-9) in order of their importance to our investigation of our galaxy. Craft one additional question of your own and bring your ranked list to class. We will then rank them as a class and begin our seminar with the "best" or "most useful" question.

Important: Don't forget our field trip to the planetarium Wednesday, so bring in those permission slips!

SCIENCE MADE SIMPLE

Anyway, people have their distractions and for me it has always been numbers, how they relate to each other, how they describe things in the world around us. Those thoughts keep expanding and unfolding until there are really cool things to think about.

I remember hearing in Earth Science class that the Earth is approximately ninety-three million miles away from the sun. That's a pretty big number: 93,000,000. It's roughly the 2009 population of the Philippines, or the population of the United States around 1910. If you started counting, and averaged one number per second, it would take you about three years without stopping to reach ninety-three million. If you subtracted out the practical time you would need to sleep and eat and all that, it would take maybe nine years. It's a big number!

But Earth is only the third planet from our sun and the distances get very large very quickly as you move outward. Jupiter's orbit has an average distance of 483,500,000 miles and Pluto's is an astounding 3,660,000,000 miles away. But even then our entire solar system is only a very small part of the Milky Way galaxy. I remember reading that it would take a beam of light one hundred thousand years to travel across our galaxy. Think about that! Light travels at one hundred eighty-six thousand miles in a single second. One second! And the Milky Way is only a very small part of the universe as we know it!

The numbers involved in such "big" thoughts get so huge so fast that scientists had to create a way of writing them in order to save time, ink and space. It's called scientific notation and it works by multiplying by a power of ten. For example, the number 93,000,000 (ninety-three million) is really just a nine and a three with six zeroes for holding place value. So this could be thought of as 93 times 6 slots of ten, ($93 \times 10 \times 10 \times 10 \times 10 \times 10 \times 10$) or mathematically: 93×10^6, where the little six, or exponent, represents the six slots. Standard scientific notation follows the pattern I described but typically uses a decimal: 9.3×10^7, which represents the same number. So, instead of writing out really huge numbers, we can simplify them using scientific notation, which saves time, space, paper and plenty of zeroes.

In astronomy, distances get so huge that astronomers have to use units of measure much larger than miles. They are called Astronomical Units (AU for short) and each unit represents the average distance between the Earth and the Sun—approximately 92,955,807 miles, or 9.2955×10^7 miles in scientific notation. These units are useful for local solar distances, for example, in case you want to book a flight to Pluto someday soon. The distance ranges from 30 to 49 AU depending on where Pluto is in its unusual elliptical orbit.

Distances to other stars require units far larger than Astronomical Units, since otherwise we would have mind-bogglingly large numbers. As I have been trying to demonstrate, using the appropriate units makes the size of the numbers more manageable and comprehensible. Otherwise, it would be like going on a cross-country trip and changing the odometer in your car to inches. The inches would pile up really fast and balloon into gigantic numbers. A trip from New York City to Los Angeles, for example, would be something like 155,992,320 inches, or in scientific notation, 1.5599×10^8 inches.

Using more appropriate units, we can instead use the easier 2462 miles. Even better, we could create our own unit, like Kelsie once suggested, and call the entire distance 1 CLAYN (Kelsie's anagram of NYC and LA), or 1 CL for short. So the distance from New York City to Moscow would be the more manageable 1.89 CL, and to Tokyo, approximately 2.74 CL, depending on the route.

By the way, is it a coincidence that the number of astronomical units in a light-year (63,240) just happens to be almost the same as inches in a mile (63,360)?

Speaking of coincidences and space and all that, an interesting one has to do with Halley's Comet, which has a periodicity of 75 to 76 years, meaning it returns to the inner solar system on a regular basis every seventy-five to seventy-six years. The interesting part is that Mark Twain was born in the same month when the comet passed in 1835, and he died the same month it passed by the next time in 1910.

He even said something like: *I came in with Halley's Comet in 1835. It is coming again next year, and I expect to go out with it. It will be the greatest disappointment of my life if I don't go out with Halley's Comet.*

Venturing out beyond our solar system requires extremely large units, starting with light-years. A light-year, or the distance light can travel in a vacuum in a Julian year, is equal to about 5,878,630,000,000 miles, or 5.879×10^{12} miles. Again, such units can bring extreme distances down to more manageable sizes. Instead of saying the star Betelgeuse, for example, is 3.76×10^{15} miles away, we can instead write the approximate equivalent 640 light-years.

A light-year might seem like a large enough unit to deal with the universe, but the known universe is incredibly, over-whelmingly huge. And I should mention gigantic, vast, enormous, mammoth gargantuan, colossal and just stunningly wicked big.

There is a larger unit of measure called a parsec (pc for short), which is equal to 3.26 light years. So, the nearest star, Proxima Centauri, is only 1.29 parsecs away, which doesn't seem that far except that the distance is roughly equal to 21,659,723,842,678 miles!

But even parsecs aren't nearly large enough. Not by a long shot. Larger still are kiloparsecs (kpc) used to measure things like the entire Milky Way Galaxy, which is about 30 kpc across, or the distance to the Andromeda Galaxy, which is a bit less than 800 kpc away. Then there are megaparsecs (Mpc), one million parsecs, and gigaparsecs (Gpc), or one billion parsecs! The Virgo Cluster, for example, is 18 Mpc away, and apparently the boundary of the observable universe has a radius of something like 14 Gpc. This doesn't sound like much, but it is something like 233,258,564,459,610,000,000,000 miles!

Of course, when your mind wanders out into space like that, it seems to be inevitable that you wonder when or if it ends at all. And that's the fun part. If the universe is infinite, then it never ends. Never! It's hard to fathom for us, but never is never. And if it does end, then what's *at* the end? If there's a barrier or something, what's on the other side? Does it wrap back on itself? Does it keep expanding? If so, what's it expanding *into*?

Some people say I "space out" all the time, but I actually like it because I think of new things, make new connections and "play" in the realm of thought. I don't know why people enjoy snapping me out of my little revelries, but they do. I find myself "spacing out" and the next thing I know someone is annoyingly snapping their fingers in front of my face, "Earth to Matt, come in Matt" or "Ground control to Major Matt, come in." One of these days I'm going to daydream myself within inches of curing cancer or finding cold fusion, and someone will snap me out of it before I can see everything clearly.

I discovered my favorite number during one of my little mind excursions. I was in history class and we were learning about the Byzantine Empire. Here's a fact we had to learn for the test: *In 313 A.D. the emperor Constantine issued the Edict of Milan, making Christianity the official religion.* As I was committing this to memory, I realized how ninja the number looked and felt when I wrote it. It reminded me of two people behind a wall talking to each other, or two people standing in line. It's also really interesting written out in Roman numerals: CCCXIII.

313 is a palindromic number, which means it can be read forwards or backwards and still be the same number. It also happens to be a palindromic number written in binary: 100111001. Most people think language palindromes are more interesting, things like "racecar" or "A man—a plan—a canal—Panama!" But numbers can be interesting in the same way. Just try finding a dollar bill with a palindromic serial number. It's not easy! Plus, you'll be looking at bills more carefully and you might just see all the other cool stuff, like the repetition of the number thirteen on the back of the one-dollar bill.

As I looked more into the number 313 (*obsessed* about the number, according to my friend, John), I found a lot of reasons for liking it. Numbers often have a lot of things attached to them and it becomes very interesting to find out more about your favorite number. Collect examples whenever you find them, and you will learn a lot about their personalities.

Anyway, for starters—literally—I was born on the three hundred thirteenth day of the year—that is November ninth. Three hundred thirteen is the number of days that the Russian cosmonaut Sergei Krikalev spent aboard the space station. 313 is also the area code for Detroit, Michigan, which is where my first girlfriend was from.

313 is a prime number, but it's also a Pythagorean prime, meaning that it can be the hypotenuse of a right triangle. Math students everywhere have to learn the Pythagorean Theorem, which states that $a^2 + b^2 = c^2$. This means that the two shorter sides squared add up to equal the longer side squared. In this particular case, $25^2 + 312^2 = 313^2$.

313 is also something called a *happy number*—yeah! Happy numbers are those that go through a specific mathematical process that ends in the digit 1. The process is to take a positive integer and replace the number with the sum of the squares of the digits. Repeat the process until it either ends in 1 or the number repeats in a cycle. If the cycle repeats, the number is called an *unhappy* number. Here's an example using the number 5:

$$5^2 = 25$$
$$25 = 2^2 + 5^2 = 4 + 25 = 29$$
$$29 = 2^2 + 9^2 = 4 + 81 = 85$$
$$85 = 8^2 + 5^2 = 64 + 25 = 89$$
$$89 = 8^2 + 9^2 = 64 + 81 = 145$$
$$145 = 1^2 + 4^2 + 5^2 = 1 + 16 + 25 = 42$$
$$42 = 4^2 + 2^2 = 16 + 4 = 20$$
$$20 = 2^2 + 0^2 = 4$$
$$4 = 4^2 = 16$$
$$16 = 1^2 + 6^2 = 1 + 36 = 37$$
$$37 = 3^2 + 7^2 = 9 + 49 = 58$$
$$58 = 5^2 + 8^2 = 25 + 64 = 89 = \text{UNHAPPY}$$

And we're back to the pattern. What's amazing is that the process of summing the squares of the digits *always* produces either a happy or unhappy number. In other words, all positive integers either fall into a repeating cycle or terminate at 1. At some point, unhappy numbers, even large ones over 1000 will plummet into the regularity of the 4 – 16 – 37 – 58 – 89 – 145 – 42 – 20 depression vortex. The only difference is where they enter into the vicious cycle and get sucked in forever, just like light is sucked into a black hole. It's hard to believe but it's true.

313 is a happy number, though, because it does not fall into that spinning-vortex-of-unhappy-doom cycle. Instead, the process reduces it down to 1, and one squared is still 1, so the process ends.

$$313 = 3^2 + 1^2 + 3^2 = 9 + 1 + 9 = 19$$
$$19 = 1^2 + 9^2 = 1 + 81 = 82$$
$$82 = 8^2 + 2^2 = 64 + 4 = 68$$
$$68 = 6^2 + 8^2 = 36 + 64 = 100$$
$$100 = 1^2 + 0^2 + 0^2 = 1 + 0 + 0 = 1$$
$$1^2 = 1 = \text{HAPPY}$$

You can work out if your favorite number, your age, address, phone number, zip code, birth date, or any number for that matter, is happy or sad. 13 is a happy number, for example, which is confusing because of the fear-of-the-number-13 thing, but it might at least explain the happiness associated with turning 13 and becoming a teenager for the first and only time.

There are a lot of other patterns to explore around happy and unhappy numbers. One pattern is that certain "number families" will all be happy or unhappy. For example, 5 is unhappy, but so are 50 and 500 and 5000 because all the zeroes squared will not affect the cycle. Another is that because the digits themselves are being used to create the sum of the squares, their order doesn't matter. So, if 313 is a happy number, then so are 133 and 331. If 39 is unhappy, then 93 will be too. Rearranging the order of the digits does not affect whether a number is happy or unhappy.

By the way, 313 is also a happy *prime*, making it even cooler.

A lot of people have favorite or lucky numbers such as 7 or 4 or 21, but most people don't have much of a reason. People often choose their numbers from their birthday, the jersey number of a favorite sports player or even something from a video game, but the number itself is only an association with something else. And I really doubt it's actually lucky for them. Once in a while I have come across someone who actually has what I consider a good reason.

My friend Kelsie's favorite number is 61. A few years ago her dad went to the hospital not feeling well. After a few tests, it turned out he had advanced cancer and they gave him only 60 days to live. Kelsie prayed all the time for him to make it past what became a dreadful number for her. To her thinking, if he simply made it past that sixty-day death sentence, then he would recover. His health declined for a while, but then he got remarkably stronger. On the 61st day, he was actually discharged. The tumor had disappeared, leaving the hospital staff completely dumbfounded.

Sixty-one has been her favorite number ever since, and that's why she changed the spelling of her name. When she heard about our system for passing notes, she added up the letters of her name. Kelsey, which is her birth certificate name, adds up to 77, but Kelsie is exactly 61. When I asked her about it, she said she wanted a constant reminder of the power of hope and belief. When she turns eighteen, she's going to get her name officially changed.

Sixty-one is a cool number. In the show *Jeopardy!* there are a total of sixty-one questions: thirty in each of the main rounds followed by the *Final Jeopardy!* question. By the way, the show's host, Alex Trebek, supposedly made his television debut in 1961. Bob Dylan has an album called *Highway 61 Revisited* and a standard eye chart, called a Snellen chart, has sixty-one letters. I haven't told Kelsie, but 61 happens to be an unhappy number:

$$61 = 6^2 + 1^2 = 36 + 1 = 37$$
$$37 = 3^2 + 7^2 = 9 + 49 = 58$$
$$58 = 5^2 + 8^2 = 25 + 64 = 89$$
$$89 = 8^2 + 9^2 = 64 + 81 = 145$$
$$145 = 1^2 + 4^2 + 5^2 = 1 + 16 + 25 = 42$$
$$42 = 4^2 + 2^2 = 16 + 4 = 20$$
$$20 = 2^2 + 0^2 = 4$$
$$4 = 4^2 = 16$$
$$16 = 1^2 + 6^2 = 1 + 36 = 37 = \text{UNHAPPY}$$

Okay, back to the code-breaking system. With the number of combinations that we needed to try, it was important not to repeat any. So I developed a simple tracking system. We started with numbers, then numbers and letter combinations and then straight letters. I hid a small notebook under one of the sinks. We took turns opening up the panel and trying a few more combinations whenever we could. So, on a given day we might stroll down to the bathroom, open the notebook, jam a paper clip in and try the next in the series. For example, 121, 122, 123, and so on. Or, A11, A12, A13. Sometimes we could get in five or six tries before being interrupted and other times none at all, because of a steady stream of unwanted visitors. Our average was 71.4 tries a day, basically what I had originally estimated.

So the six of us were headed to the bathroom whenever we wanted. The teachers rebelled a little at first, but given the doctors' notes and the strange circumstances of all of us having eaten weird things the same night, they finally gave in. I said "six" because nearly every day when I checked the notebook, there were a number of entries written in Kelsie's unmistakable handwriting. I don't know how she pulled it off, and whenever I asked her, she only ever smiled. She's amazing sometimes.

Actually, she's amazing all of the time.

Of course, the original doctors' notes only lasted a few weeks, but with a few twists of the details, some good acting (in Kelsie's case, some great acting), we extended our bathroom visits for several weeks. After that, it was a lot of waiting.

5

Waiting

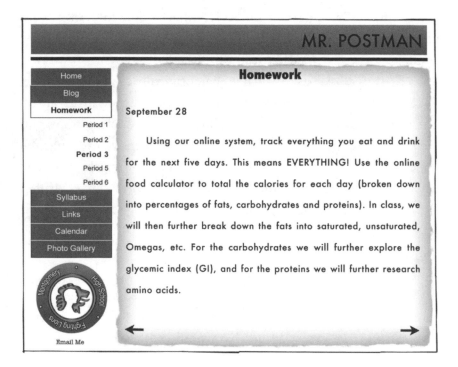

School lunches unify students throughout the country because of how universally terrible they are. I think they must be close to illegal, like grade F beef or something that would cause mad cow disease even in a mad cow. Now that our school was renting space, the lunches were actually made at the middle school and delivered to our new building, where they were reheated and then sold. If you have bad school lunches in your school, just imagine them as reheated leftovers.

Lunchtime is a strange thing at schools. They cram us into a relatively small space, feed us terrible food and expect us to be quiet about it.

Then they send around angry lunch ladies who are always mad at the world and ready to yell at nearly anything. Jason Stone swears he got yelled at for using a spoon instead of a fork, and Scott Andrews said that he once got yelled at for standing up too soon before lunch ended. Exaggerations maybe, but I think if lunch ladies were armed with Billy clubs, there would be a lot of injuries for the nurses to handle.

Lunch was the only period we all had together. We sat around a circular table and in the beginning talked about how the system was going and whether we could possibly speed things up. Sometimes we speculated about what was going to happen when we cracked the code. Jamie insisted that nothing much was going to happen and that we were just wasting our time. John thought it was a face recognition program or something like that. Thomas and Ari had a theory that the whole thing was some kind of secret training program.

By early October our lunch conversations about the strange computer dwindled. We had spent so many hours speculating and dreaming about winning the lottery or becoming famous by cracking this code, that we had exhausted our possibilities. Day after day passed with no tangible results, and, although we kept the system going, we stopped talking about it openly.

My classes were going fine, but I had already tracked every teacher's use of ums and ahs (no new winners this year), had calculated all of the windows for asking to go to the bathroom. I started graphing the speaking volume of certain teachers, mainly because of Mr. Haines. He was a good lecturer and really knew what he was talking about, but for some reason his voice dropped in volume the farther along in a sentence he got. He boldly started every sentence with power and volume as if he was addressing an auditorium full of people. Then he would decline from there until the end of every sentence was barely audible.

Lunch started to become my favorite time at school, mainly because we were all together, but also because of one of my new endeavors:

probability. This branch of math is completely ninja, mainly because it messes with your head—at least it messes with mine.

The birthday paradox is a great example of something that seems like it shouldn't work. Here's the idea: How many people would there need to be in a room, so that there are at least two people with the same birthday? Of course, you might answer 366 to make sure that every day of the calendar is covered. But that's not the weird part. As it turns out, if there are only 23 people in the room, there is a greater than a fifty percent chance of two people sharing a birthday! So, in many classrooms at school there actually should be two people with the same birthday in the room. With forty people, the probability increases to an astounding ninety percent, which is the part that messes with me.

Probabilities only describe what is *probable*, not what will actually be, and they are typically expressed in the form of fractions. The top number, or numerator, represents the number of outcomes that are desired or that you want to know. The bottom number, or denominator, is the total number of possible outcomes. Take flipping a coin in this case. What is the probability that I will get heads? Heads is the desired outcome, so the numerator is 1. There are two total possible outcomes, so the denominator is 2. The chance of getting heads is then expressed as the fraction: ½. So, flipping a coin should result in half of the outcomes being heads. But that is only *probable*. You could flip the coin a thousand times and *could* end up getting ALL tails … it's just very *improbable* that it would happen.

For someone like me who loves math for the surety, for knowing that there are specific answers for most problems, probability makes me insecure. I know that the probability of the coin flip is ½, but that doesn't guarantee anything. If I flip a coin and get heads, the probability of the next flip is still ½. If I flip it again and get another heads, the probability of the next flip is still ½. That really bothers me. I want to *know* the next result, not just predict it. Of course, if I could predict things like that, then I would win the lottery. Actually, I'd win every lottery eventually.

If that isn't bothersome enough, I once heard someone say, "All probabilities are fifty percent: they either happen or they don't." At first I laughed, but then I started thinking about it more and more. Now it just scares me because I can't shake the idea. I feel like half the time it's a great joke and the other half it's totally right.

Thinking about flipping coins and probability just raises more unanswered questions: Does the flip itself matter? Are coins lopsided, and therefore, produce different results? Could a coin ever land on its edge? I saw an old black and white *Twilight Zone* episode where that actually happened, but I've never been able to get a coin to land on its edge.

Anyway, at lunch I started tracking probabilities, partly for something to do and partly to try and understand things better. Plus, it's remarkably fun. Here are some of the first things I observed. The probability of having corndogs for lunch is 1/10. The probability that one of the lunch ladies will yell is on the chart below. When I say *yell*, I don't just mean that they raise their voices over the noise in the cafeteria. What I mean is a horrible screeching sound directed at certain people for minor things, like the time John got yelled at for pretending to be a mime. Come on! A mime!

Mondays:	1 in 3
Tuesdays:	10 in 25, or 1 in 2.5
Wednesdays:	1 in 5
Thursdays:	1 in 2
Fridays:	1 (That's 100%. They've yelled every Friday).

The probability that something unidentified will be in the gelatin dessert is ¼. After dissecting numerous gelatin desserts, we have yet to understand what some of the things we found actually were. Hairs in the gelatin stopped being unusual or interesting years ago, but there have been some really weird things. We began the Gelatin Olympics by starting a rating scale for the weirdest objects ever found. Every month we gave out prizes for the best ones. The scale was:

 0 Absolutely nothing unusual
 2 Hair or other explainable object
 4 Unusual explainable object, such as a part of a spoon
 6 Unusual object
 8 Unexplainable object
10 Completely random and weird object

Over the past few months we found a lot of strange stuff, like when Kelsie found a feather in some green gelatin. She doesn't actually eat the school lunches like I do, but she does like to go for gold, so she sometimes walks around the cafeteria and inspects everyone's gelatin desserts. The highest rated object we ever found, with an average rating from our table of 9.8, was a thimble. Yes, a thimble. Lisa Rider found it in a rectangular prism of orange gelatin. We came up with several theories about how a thimble would end up in a school lunch, and the current favorite is that the cooks were playing Monopoly while waiting for some of the food to finish cooking and the playing piece ended up in the gelatin. We asked around if anyone found a small car, a dog, a top hat or one of the other pieces, but no one had.

The best object I ever found was when I got what I swear was a rat's foot, though Kelsie insisted it was just part of a banana peel. John and Ari believed it was a grape stem. I couldn't prove exactly what it was, so unfortunately, I didn't get any points. I'd like to think it was a rat's foot, since it would have probably been worth a gold medal that month. I didn't eat the mystery object in the gelatin, by the way, but Corey did.

What would lunch be without our school-famous Corey Meunch, or Muncher as we called him? From what I've heard, every school has someone like Corey. He's the kid who will eat anything or do just about anything for a few bucks so that he can get an extra ice cream or cake at lunch.

Here's Corey's normal operation. Take the leftovers from a whole table, mix them together into a remarkably gross concoction, and Corey will eat it for certain amounts of money, which I started tracking weeks ago. So, take the gelatin from a whole table, mix it with both regular and chocolate milk, throw in some corndog bits, some wilted lettuce and Italian dressing, and he's likely to eat it.

How likely? Well, I can tell you. It's based on how much he'll get paid for his feat.

Any concoction, $2 or more: 1 (100%)
Any concoction, $1-$2:   3 in 4
Any concoction, under $1:   2 in 3

Corey has other talents as well. He once licked a path down the floor of the main hall for $20. I was there and still remember seeing the clean path that he created on the floor tiles. When he was done, he smiled and stuck his tongue out at us. It was covered in dust and grime. My friend Lauren almost puked, and I have to admit that I almost did too. There comes a point where gross becomes disgusting.

It's also rumored that Corey licked the inside of a urinal for $25, but no one I know was actually there to see it.

Then we have Marshall Morris. I feel bad for him sometimes. He was a friend of mine in elementary school, but somehow we just grew apart and now I almost never talk to him. Now, I'm not trying to be mean here, but I simply observed a specific thing about him. At any given lunch period, the probability that Morris will pick his nose is 1 in 3. The probability that he will eat it too is 1 in 5.

Here are some other approximate probabilities I have discovered while sitting around at lunch:

Probability that I can make Lauren puke, normally:	1 in 8
Probability that I can make Lauren puke, watching Morris:	1 in 4
Probability that a food fight will break out, large:	1 in 37
Probability that a food fight will break out, small:	1 in 12
Probability that Jessica will wear all black, socks excluded:	1
Probability that Jessica will wear all black, socks included:	1 in 4

I also collected some from other classes:

Probability that Neil will get detention, math class:	1 in 3
Probability that Neil will get detention, other classes:	1 in 9
Probability that Mr. Haines will get overly excited and accidentally break something during class:	1 in 5
Probability student will get hurt in gym class, dodge ball:	1 in 2
Probability student will get hurt in gym class, floor hockey:	1 in 3
Probability student will get hurt in gym class, other:	1 in 8
Probability of seeing blood spilled in class, gym:	1 in 14
Probability of seeing blood spilled in class, other:	1 in 92
Probability Dan gets hit in the face by playground ball:	1 in 3
Probability anyone else gets hit in the face by playground ball:	1 in 12

I've thought about going to the superintendent about some of the teachers, particularly Ms. Simmons, whose probability of being boring is 100 percent. But I don't know what the superintendent can do about boring teachers. I imagine you can't get fired for being boring. Besides, I'm not sure who's at fault if something is boring. Think about it: if I find math fascinating and you find it boring, then math itself can't be the thing that's boring. Or if I find hockey boring to watch and my dad doesn't, then it can't be hockey that's to blame.

All this leaves a particularly unsettling feeling in me because it leads me to only two main conclusions: Either Ms. Simmons actually is boring, or I am bored in her class. It looks like Ms. Simmons is animated enough and seems to love what she's saying. That can only mean that I AM bored, at least in her class. Put another way: I am bored in her class, *I get bored in her class or boredom assaults me in her class* The point is that I realized that I was responsible for the boredom, NOT her!

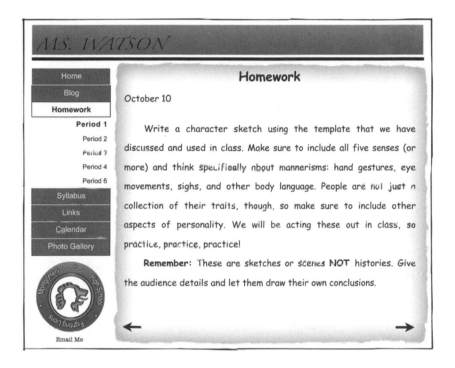

MS. WATSON

Home	**Homework**
Blog	October 10
Homework	
Period 1	Write a character sketch using the template that we have
Period 2	discussed and used in class. Make sure to include all five senses (or
Period 2	more) and think specifically about mannerisms: hand gestures, eye
Period 4	movements, sighs, and other body language. People are not just a
Period 6	collection of their traits, though, so make sure to include other
Syllabus	aspects of personality. We will be acting these out in class, so
Links	practice, practice, practice!
Calendar	**Remember:** These are sketches or scenes **NOT** histories. Give
Photo Gallery	the audience details and let them draw their own conclusions.

Email Me

← →

As fun as collecting data during lunch was, I discovered a whole new territory when we had a run of substitute teachers in the early weeks of November. I don't know what you need to do to become a sub, but I imagine that it can't be much. Now, we've had some great subs in the past, and some of them really know what they're talking about, but many of them are there simply to supervise us.

I know subbing must be hard. You have to step into a class and try to teach when you don't know the students, and often don't know the material. You almost certainly have no context, even if you do. And student behavior is typically atrocious. Even considering all those things, some of the subs are so indifferent that they don't even try. They read the newspaper, talk on their cell phones and check their email. They could at least *try* to teach us.

Anyway, subs became a new favorite of mine, not because we learned anything, but because they were so interesting. Here are some examples:

Probability Mr. Kelly will smell like cigarettes, morning:	1 in 3
Probability Mr. Kelly will smell like cigarettes, afternoon:	1 in 2
Probability Ms. Shaw will tell wildly random story:	1 in 2
Probability Mr. Dent will say the phrase "kids these days":	1
Probability that sub can actually help us in math class:	1 in 6
Probability Ms. Andrews will shop for shoes online:	1
Probability Mr. Deally will give someone detention:	1
Probability Ms. Bond will give detention:	0
Probability Mr. Washburn remembers your name, front rows:	1 in 2
Probability Mr. Washburn remembers your name, back rows:	1 in 7
Probability Mr. Evangeline will audibly pass gas, before lunch:	1 in 4
Probability Mr. Evangeline will audibly pass gas, after lunch:	1 in 2
Probability Ms. Harris will have lipstick on teeth, morning:	1 in 4
Probability Ms. Harris will have lipstick on teeth, afternoon:	1 in 3
Probability we can convince sub that our school has no tests:	2 in 7
Probability we can convince sub that homework is optional:	1 in 3

6

Discovery

On Discovery Day, or D-Day as we got to calling it, there were four of us there: Jamie, Thomas, John and me. According to my little code-breaking log, we punched in JAZ. After watching the screen revert back to the three blank boxes hundreds of times, we fully expected it to happen again. But this time the screen went black for an instant and was replaced by a light blue background with just a cursor blinking in the top left corner.

We heard a popping noise over by the third stall, the one that doesn't work. We hurried over and ... let me pause here. If you ask most kids to make a list of the ten coolest things they could think of, you'd probably get a pretty similar list: becoming a sports star, winning the lottery, driving a fancy car, dating a Hollywood star, and so on. But what we saw past the third stall, in my opinion, should make most lists, because along the ridge of the bathroom wall, where the brick transitions to drywall, we saw an open secret door.

Remember that irresistible force that drove me to push those three buttons? Well, finding a secret door in your school is far worse. We were through that door and into a dark, narrow corridor before even considering whether it was a good idea or not.

Jamie was the first one through. I was just behind him with Thomas and John just behind me. Some of the light from the bathroom spilled into the corridor, but I couldn't see much since Jamie was blocking most of my view. A few steps in and I couldn't see much at all, so I put my hand on Jamie's shoulder. I thought about Thomas and John getting somehow separated and then numerous horror movies passed through my mind, ones where the characters got killed in gruesome ways because they left the group.

Maybe that was the secret moral of all of those horror movies: you always stick together no matter what. Separate and you die. In a way that's kind of true. When we're young, we are totally dependent on others to take care of us. Then, as we get older, we start becoming independent. We can range out as teenagers and, well, explore secret doors in schools. But independence is not the end. We eventually need to become interdependent. Just like the web of life in biology (yes, I do pay attention even when I have been going to the bathroom three or four times a class), people are always connected to others in mutually beneficial ways.

Anyway, none of us died. We shimmied through the corridor for what I estimated was 51 feet until we came to a door. I guess it's no surprise that it was locked—and locked with a code of some kind. We were so excited about finding a secret door in the school that we must have assumed that the door would open because we rushed forward and when it didn't open, we crashed into each other in a cartoon-like manner. Actually, it reminded me of the Three Stooges, except that there were four of us, so I guess I have to include Shemp, although it pains me to do so, since he wasn't one of the main three. By the way, in addition to Moe, Larry, Curly, and Shemp, there were actually two other stooges: Joe Besser and Curly Joe Derita.

The door was made of some metallic alloy. A three by three keypad with the numbers one through nine was centered on the door. The numbers were all black, except the number 8, which was green and the number 2, which was red. Above the grid was an embossed gold 15. Below the grid was a small, blank screen.

"Another code?" John whined.
"Seriously?" Jamie moaned.
Thomas shrugged, "Must be getting close."

"Close to detention!" John snapped. "Not to mention being grounded and who knows what else."

Jamie shook his head, "All this work just to find another code?"

"Okay, Matt," John said, "this was all your idea. What do we do now?"

"We solve this too," I said.

They backed away from the door a little to give me more room near the keypad. I guess they expected me to figure everything out right then and there. I took a deep breath.

Of course, *square* and *squared* are very close, so my mind instantly went to squared numbers. I pressed 1, 4, and 9 in that order, but I guess that would have been too easy. The lock didn't open. Just pressing those buttons didn't take into account the number fifteen above the keypad or the fact that the two was red and the eight was green, so it's no wonder the door didn't open.

Not using all of the information, or at least taking it into account, is a classic issue with solving word problems. I have seen a lot of other students get questions wrong because they only used some of the information presented in a problem. They might read the first sentence and feel that they have what they need in order to answer. But this is rarely true. Word problems often require *combining* information together or using multiple steps in order to arrive at a complete answer. Trying to answer a word problem without taking into account all of the information, is like trying to guess the size of a house by entering one room.

I started talking to myself, "So, how does the number fifteen relate to the keypad?"

We studied the puzzle for a moment and then Jamie jumped in, "The second row adds up to fifteen."

"And the diagonal," John continued, "wait … actually both diagonals add up to fifteen."

"The middle column, too," Thomas added.

John was getting excited. "So, press one of those sets. No, wait, dude, press the middle column but go up because the eight is green and the two is red. Get it? Green is go and red is stop!"

I had to admit, his idea sounded reasonable, although maybe a bit too obvious. I was still thinking about it when he reached over and pressed the buttons in that order: 8-5-2. Nothing happened. "Ah, dude," he moaned, "I thought I had it."

"Wait," Thomas leaned in, "maybe it's like an equation! Eight plus five plus two equals fifteen. So we have to punch the one and five at the end."

John motioned him to try and Thomas pressed 8-5-2-1-5. The blank screen became a little brighter and then a message scrolled across:

DOOR HAS BEEN RESET FOR 24 HOURS

"Ah crap," John and Thomas said nearly in unison.

Jamie shook his head, "I knew it."

"Knew what?" John said looking aggravated.

"I knew those codes weren't right."

Thomas looked mad too. "No you didn't! You didn't say anything because you didn't know either."

Jamie just walked away mumbling to himself and looking dejected. You would have thought he had just lost the Olympics or something.

"Matt, what's up with him?" John asked.

I shrugged, "I think this sneaking around school is really getting to him."

"Whatever," was John's reply. "Come on," he said, "I guess we should get back to class. He and Thomas shimmied back down the hall and then went through the secret door into the bathroom. I took one last look at the keypad and realized I had the start of an idea.

As I mentioned, some people like to play with clay because they can mold and shape it into creative forms. Some people like to play with words by writing poetry, fiction and plays. Some people like to play with food by trying new recipes. Others like to invent games and play sports. The point I'm making is that people like to play with stuff. Not many people enjoy playing with numbers, though—once you rule out those interested in money, that is.

Just look at someone's notes in class and you can probably tell what they think about. I'm talking about doodling. I bet it would be fascinating to study doodling considering that people are absent-mindedly drawing little pictures of what goes unfiltered through their minds. Anyone who looks at my notes will know instantly that I love playing with numbers. I write numbers upside down, sideways, backwards, and sometimes mirrored. It's weird, but if you write a number enough times it starts to feel strange and cool things can emerge. An obvious one is that six flipped upside down looks like a nine. Less obvious is that two overlapping eights at ninety-degree angles look like a four-leafed clover.

Doodling with numbers raises interesting questions. Why do people draw numbers in different ways? Why do some people draw a line through their sevens? Why do some people draw their twos with a loop? Why do some people draw their eights with a single twisting stroke, while others draw two separate circles? Once in a while someone will draw their eights by connecting a three and a backwards three. It's unusual, but I've seen it.

I was doodling in class the next day when I thought more about the keypad on the door. As John mentioned, the colors were the first clue. I had to admit, green meaning go and red meaning stop made a lot of sense. I still liked that idea. The gold fifteen was the next clue. As we had mentioned, there were some combinations of rows, diagonals or columns that added up to fifteen. But they weren't ALL fifteen. The first row, for example, only had a sum of six. What I found interesting while doodling, though, was that the numbers 1 to 9 add up to a total of 45, and average 5. This made sense, since, as we had noticed, three of the numbers together can add up to 15. For example, $9 + 2 + 4$, or $8 + 4 + 3$.

That's when I realized that if the numbers were arranged properly, they could all add up to fifteen. In every direction: the columns, the rows and the diagonals. All of them. I realized, then, that the problem we faced was something called a magic square.

The first time I saw a magic square was actually in art class. It was in a piece by Albrecht Dürer entitled *Melencolia I*, an engraving from 1514 with enough symbolism to talk about for months. Here's the actual magic square from the engraving:

16	3	2	13
5	10	11	8
9	6	7	12
4	15	14	1

Notice that all of the columns, rows and diagonals add up to 34. What's even more amazing is that all four corner quadrants also add up to thirty-four. Plus, the middle four squares add up to thirty-four as well, since $10 + 11 + 6 + 7 = 34$. Two last things: the middle two bottom squares mark the date of the engraving, 1514; and the bottom corners marked 1 and 4 supposedly correspond to A and D in the alphabet—not only Albrecht Dürer's initials, but also symbolizing A.D. for the year. So magically the bottom row represents 1514 A.D. Ninja cool!

As cool as Dürer's magic square is, though, my favorite one is a little more complex, since it stays "magic" even when flipped upside down!

96	11	89	68
88	69	91	16
61	86	18	99
19	98	66	81

As far as the keypad was concerned, the numbers could be re-arranged to form a magic square. I started with the eight in the top left corner and ended with the two in the bottom right corner. Since each row, column and diagonal needed to add up to fifteen, the missing middle number was $15 - 8 - 2$, or 5. So, here's what I had so far:

8		
	5	
		2

If this premise were true, then the boxes to the right of the eight and the boxes below the eight would each have to add up to seven, since 15 − 8 = 7. The remaining two larger digits, 7 and 9 would have to go in the other boxes. That really left two options: either the 7 went in the bottom square under the five, or else the 9 did. Once those two numbers were determined, everything else falls into place. If the 9 went in the bottom box then the bottom row would be 9 + 2 so far, leaving a 4 in the bottom corner. Then the first column would be 15 − 8 − 4, leaving a 3. The middle row would then be 3 -5 -7. The middle column could be determined by 15 − 9 − 5, or 1 and the last column would be 15 − 7 − 2, or 6. So that would mean:

The 7 and 9 could be reversed to create a similar, but slightly different square:

8	1	6
3	5	7
4	9	2

8	3	4
1	5	9
6	7	2

The question in my mind was: Which square was right? I figured if we tried both, we'd be able to open the door, so I wasn't worried too much.

Next period I waited for the appropriate time and headed back to the bathroom. I made sure no one was coming, punched the sink code and hurried down the secret hallway. I typed in the numbers from the first magic square, in order from left to right, top to bottom: 8-1-6-3-5-7-4-9-2. I thought one of the codes was going to open the door, so I was surprised when the gray screen on the bottom blinked three times. A message scrolled across:

* * * INPUT PIN * * *

Then three lines appeared on the gray screen. Apparently, three numbers needed to be added as a pin code. I thought about the J-A-Z that got us past the mirror code and flipped my cell phone open to see what corresponded to those three letters: 5-2-9. I punched them in and was annoyed to find that the screen blinked again and was cleared. I probably had one or two more attempts before the whole thing reset for another twenty-four hours. Rather than try a few more guesses, I left the keypad and the hallway and went back to class.

Time for more math doodling.

I played around with the magic square a bit more and found a number of other patterns. Jamie recommended that I try to use multiplication or other operations to search for other patterns. For example, although the nine numbers add up to 45, they multiply out to 362,880. I figured that perhaps the pin number might be 362 or 880, but it didn't seem to fit.

What I did discover was fascinating. In either version of the re-arranged magic square (the rows of one are the columns of the other), the following calculations can be made:

$$
\begin{array}{ll}
8 \times 1 \times 6 = 48 & \quad 8 \times 3 \times 4 = 96 \\
3 \times 5 \times 7 = 105 & \quad 1 \times 5 \times 9 = 45 \\
\underline{4 \times 9 \times 2 = 72} & \quad \underline{6 \times 7 \times 2 = 84} \\
225 & \quad 225
\end{array}
$$

I even found a similar pattern with the diagonals by multiplying the digits and adding them together. Then squaring the middle digit "5", the total is once again 225.

$$
\begin{array}{r}
8 \times 5 \times 2 = 80 \\
6 \times 5 \times 4 = 120 \\
5^2 = 25 \\
\hline
225
\end{array}
$$

Maybe this was the three-digit number we needed.

7

Beyond the Door

After school I gathered up everyone except Kelsie, who had band practice. We shimmied down the secret corridor and came to the keypad. After punching in the numbers again, the pin lines appeared.

"Matt, you have a plan, right?" John asked.

"Yeah, I worked out a bunch of patterns and 225 kept coming up. I'm going to try it," I said as I entered the numbers. The door clicked open slightly.

"Yes!" Jamie and Ari yelled. Thomas looked happy, but John seemed strangely unimpressed that the door was finally opened.

"Well, duh, 225 makes sense," he said. "Didn't you say that the columns and stuff add up to 15?"

"Yeah," I said pulling the door open further with my fingertips.

"Well, 15 squared is 225."

I felt like smacking myself. After pages and pages of mathematical doodles (many of them I considered semi-professional grade, with a few cartoon characters thrown in for fun), I had painstakingly arrived at the number 225—only to have John come to a similar solution in his head in like ten seconds. Oh well, at least we got the door opened.

When I opened the door, I saw a large room full of shelves and huge machines. You know those mad scientist laboratories in the movies that have all kinds of crazy gadgets and glassware? Well, this was clearly one of those rooms—except not so much with the diabolical glassware. Almost every square foot of the room (close to 3000 square feet since the room appeared to be about 60 feet long by 50 feet wide) was packed with stuff. Small aisles allowed movement through the place, and everywhere there was mad scientist stuff: tables overflowing with books and diagrams, computers with wires interconnecting into a small mass, chairs piled with electronic parts, bins full of what looked to me like miscellaneous generators, motors, batteries and tools, and everywhere dust. Nobody had been in here in quite some time.

It occurred to me that maybe this was an old classroom that had been abandoned or sealed off for some reason. Maybe a partition had been erected at some point to make use of other space in the building and this either got lost in the shuffle or ... that was when I looked closer at the dust on everything. White powder. That was when I got scared. Maybe this room had been *quarantined*.

Quarantine is an interesting word. I read somewhere that it essentially means "forty." I think the idea is that forty is somehow a natural period of time to ensure that something is cleansed or purged. Maybe that's because forty is ten times four and the number four represents a cycle. Think about spring, summer, fall, winter, or earth, air, fire and water. Or even think about night, morning, afternoon, evening. In cards there are four suits: hearts, spades, clubs and diamonds.

So maybe ten cycles of four and something is completely purged. Maybe it's nine cycles (nine being the last single digit and an end in that sense) and then one more for insurance. It makes me think about octaves and music with seven notes plus one as a return to the beginning in a sense, only this is with ten. If you think about the number 10, it's really just a "1" in the tens place. Maybe ten cycles because humans have ten fingers and ten toes. Or maybe because ten is the first double digit number and in that sense is the beginning of numeric octave.

Anyway, quarantine has to do with forty. It's actually fascinating since the number forty comes up in a lot of interesting situations. Take the Bible, for example. Jesus fasted in the desert for forty days and nights. In the flood story, it rained for forty days and nights. The Israelites wandered for forty years. Moses was on the mountain for forty days and nights. And so on. In each of these instances, the idea of quarantine or purging or cleansing works.

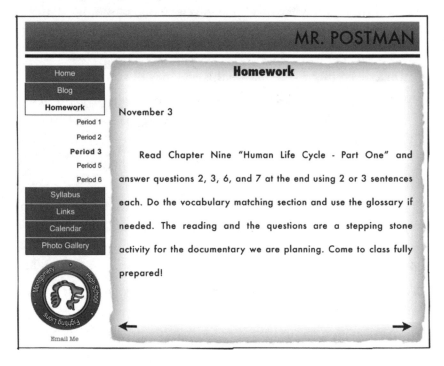

On the flip side, forty can also be a process of building or growth. Take, for example, an average human pregnancy, which is forty weeks. This isn't quarantine in the purging sense, although I can still see it in terms of the woman's body during pregnancy. For forty weeks she must (or should) keep herself as clean as possible so that the baby is not affected by nicotine or alcohol, for example. There's also the expression "sleeping for forty winks," which is a quick nap to refresh the mind and body.

I had a moment of panic about the white powder since a few years ago there was all that Anthrax scare, but I figured that there would be no possible way a school could open up in this building if that were true. We had no way of finding out in any case—unless, of course, we started dying horrible deaths. It's not like we could just bring some of the dust up to the chemistry lab and ask Mr. Henry if we could research this strange white powder that we found while skipping class.

As I was pondering the meaning of forty and all that, the bell rang, which meant we had to hurry. Not only did we have to shimmy back through the corridor, but we had to make sure that the secret door was closed, that we shut down the sink computer, got our books from Art class and tried to make it to social studies in time.

We had two weeks left for our current run of doctors' notes, so we spent the time accessing the mad scientist lab that we started calling Doctor Frankincense's Lab because of the strange odor that came from the heaps of old books that littered one of the back corners.

The procedure was basically the same: we got excused from class, sped down to the bathroom, opened the control panel, punched in the J-A-Z code, and then shuffled down the narrow corridor. Then we plugged in the door code. We could only spend a few minutes at a time during class, so we started staying after school, since it was usually open another two hours for after-school sports and activities. We told our parents we had a science project to work on, which was basically true.

Between sneaking a few minutes in during classes and the hours after school, we cleaned the lab in a few weeks. The hard part was all of the dust. In such a confined space, it was hard not to throw a lot of dust in the air and Thomas, in particular, had a tough time with his allergies. Jamie brought some dust masks from his father's garage, but those didn't help entirely. Our goal was to clean the place up as a first priority, especially because there were used ketchup packets scattered everywhere. As you might imagine, we stopped on numerous occasions to read books, browse at diagrams and play with some of the science toys we found.

Thomas and Jamie weren't terribly interested in the lab equipment because they said it was too much like doing additional schoolwork.

From their perspective, the secret lab was a glorified lounge, and they were perfectly content to go there and hang out. Once they cleaned the place up a bit, they moved the furniture around until there was a lounge area outlined by three old leather couches. Before I knew it, they had hooked up two small refrigerators and stocked them with a variety of power drinks. Thomas even brought in a stereo and his old Wii console so they could listen to music and play games.

The most excited they got was when they opened a locked cabinet and found close to two thousand old vinyl records. They danced around and hollered as they found albums by their favorite classic rock bands. I didn't think they could play them until one day Thomas yelled out, "Whoa, guys get over here!" There was so much excitement in his voice that we all dropped what we were doing and ran over. "Check this out!" he said as he grabbed a Led Zeppelin record.

Next to their stereo was a tall rectangular box, with a metal casing on the bottom and glass on top. It looked like one of those big popcorn machines at the movie theaters. Jamie stood on the other side with a big grin on his face. Thomas slid the album into a thin slot in the side of the machine and we watched as the record got mechanically pulled in. Then it appeared at the bottom of the glass cube, started spinning and *Ramble On* blasted out the speakers in the base. The sound was incredibly clear with no static.

"Dude, that's it?" John said. "That thing's as big as a refrigerator. I mean …"

Jamie held up his hand. "First, take a closer look."

John, Ari and I leaned in. I noticed that although the record was spinning, there was nothing visible *playing* it.

"Now watch," Jamie said. He pulled a lever on the side of the machine and the album floated up in the air and started spinning around in three dimensions. It spun so fast that it looked like a bowling ball suspended in mid-air. The song kept playing without missing a beat.

"That is *so* beast," Ari said, which is what I was thinking and, based on John's face, probably what he was thinking as well.

Kelsie was suddenly standing next to me. "I was wondering what that was."

John was most intrigued by a giant donut-shaped contraption near a bank of old computers. I pointed out that such a shape is technically called a torus, but he continued to call it a donut. Anyway, it reminded me of numerous portals or gateways from science fiction shows and movies. It was about eight feet in diameter (25.13 feet in circumference—that's pi times eight or 3.14159×8), with blackened wires coiled around the edges. A thick cable snaked over to a fried electrical panel. Attached to one side of the torus was a curved, three-tiered keyboard with eight rows of five keys each, or 120 keys ($3 \times 8 \times 5$). They were numbered 1 to 118 with two blank keys at the end. I impulsively pressed a bunch of the keys, but nothing happened.

Above the donut was a sign that read CAUTION. John shrugged when we first saw it. "Well, that's a no-brainer."

"What do you mean?"

"Two things. Of course it says caution—that thing draws a crazy amount of electricity. Two: someone *didn't* use caution and fried it. One false move and we'd get fried like a plantain."

"A plantain?"

"Yeah it's a fruit that looks like a banana ..."

"I know what a plantain is! I'm just saying *like a plantain* is the best simile you can come up with?"

"What's a simile?" he said smiling. Then more seriously: "Dude, I just had some last night. They're so good!" He smacked his lips. "You ever had some?"

"We have all this cool stuff to explore, and we're talking about a plantain?"

"Dude, they're so good, though. Sprinkle a little cinnamon ..."

"John!"

"Okay, okay," he pleaded.

I wasn't surprised that John wanted to work with the torus. You should see his room. I know a lot of kids pride themselves on how amazing their rooms are, once they clean enough junk off the floor: cool posters of celebrities, cars, sports stars or whatever. They may have painted their own rooms, maybe even with one of those fancy techniques you can learn at a home supply store. They might have really stylin' furniture with special sheets on the bed, or a roll-top desk or something. They might have a loft with the bed built into the wall. But no one else I know has a train that runs through their whole house.

When I say *through,* I really mean it. His G-gauge train is set up about sixteen inches from the ceiling and runs through tunnels built in all the walls. It all started when he ran out of space in his room. His train took up so much space in his bedroom that it was basically impossible to move in there without knocking over a miniature water tower, a tree, or another scale model of something. When he was about eight, he decided to expand, went out to the garage, grabbed a maul and punched holes in the walls of his house.

At first his parents were furious, of course. But after they cooled down, I think they realized that they hadn't been spending enough time with him. Over the next year they set up the train throughout the house, so that it could travel through every room except the bathroom and his parents' room. For a few years after, we could be playing in John's room, attach a request note to the train and send it to the kitchen. After a while, we would bring the train back around and John's dad would have made sandwiches and put them in the cargo cars, or filled the tankers with soda for us. Ah, those were the days of sheer lazy play. When we got older, we had to "get off our lazy teenage butts" to get stuff from the kitchen. Oh well.

Anyway, the train is still there and more complex than ever. John has it wired to do just about anything you can imagine a toy train could do. His closet is overflowing with wires of all kinds that activate the railroad crossings in the living room, or make a tiny figure wave to the train in the kitchen or flash lights in one of the train stations out in the hallway. He even rigged it so that the train whistles when their doorbell is pressed.

His floor has bits of wire scattered around, with an assortment of pliers, wire-strippers, spools of copper wire and so on. So, you can understand his fascination with the electric gizmo. He spent days and hours tracing the electrical lines and the cables that connected to the old computers. He had a strange way of working, at least from my perspective. He sat on one of the desks, sometimes for fifteen or twenty minutes, with his face all screwed and knotted with thinking lines. Every once in a while he would leap off the desk, crawl around on the floor and connect a few cables. Then he would nod to himself and climb back on the desk. Not exactly how I liked to work, but he got the torus thing working eventually.

8

Not Your Average Journal

Ari and I were most interested in a leather-bound book that we found inside a desk drawer. We figured if we were going to find out more about this secret room and all the stuff, then the book was a good place to start. We got our hopes up thinking we had found a mad scientist's secret journal describing in detail how to reanimate dead bodies, build intergalactic space ships, or at least something cool like using nanotechnology to give us superpowers.

The book had a circle on the cover, with a diameter line parallel to the binding and nine buttons numbered 1 to 9 around the circumference. An infinity symbol was embossed on the binding. I was disappointed thinking that we had only found one of those cheap puzzle books that are common in the clearance section of every bookstore. We opened the cover and found the pages completely blank inside. I was about to complain about how boring this book was compared to my imagination, but that's when it became even more intriguing.

Ari flipped through all thirty-eight of the pages (I happened to count). By the way, thirty-eight is the sum of the squares of the first three prime numbers ($2^2 + 3^2 + 5^2$).

"Matt, these pages are like sheet metal or something."

"Whoa," I said bending one of them back and forth. "But they seem more like ..."

"Computer screens?" Ari suggested.

"They're like iPad paper or something," I said.

"This is beast! It's like digital paper or something. Hey, wait ..." Ari scratched at the binding of the book. A panel slid open in the binding revealing a compartment inside. "I guess that answers it," he said pointing inside. There were several different ports in the compartment, one was clearly a USB, one was firewire, and there were three others that we didn't recognize.

Ari pulled out a stylus much like one for a tablet computer. "Check this out," he said, unhooking something. The stylus was hinged at the top, and I watched as Ari pulled it open until it resembled a drawing compass or those measuring calipers that you always see in the old submarine movies. Ari pinched them and they sprang back to an open position.

"I've never seen anything like that," I said.

Ari snapped them together at me a few times, "They're like using chopsticks."

"So ... it's a computer?" I asked, although it wasn't really a question.

"I guess so."

"Disguised as an old puzzle book?"

"Yes, Matt," Ari said impatiently. He picked the book up and spun it around a bunch of times. "There's no power button."

"There are nine on the front. What are those for?"

"I don't know, but there's no power button."

Ari kept spinning the book around until he had looked at the whole thing up close. Then it dawned on me: the puzzle on the front was the power button! Ari looked at me and seemed to know what I was thinking. "This thing has a code just to switch it on?"

I nodded, "It looks that way."

Ari set the book down. "Here, Matt, it's all yours. I'm gonna listen to some tunes with Thomas and Jamie.

I took another look at the book: a circle embossed on the leather cover with a single line through the center created a diameter that was parallel to the binding. Nine numbered buttons were arranged around the circle, starting with the number one in the same place as a clock face and continuing clockwise around the circle until the number nine was at the very top. An infinity symbol was embossed on the binding. Other than that, there were no other clues.

I tried pushing the numbers in various obvious sequences, such as: one through nine, nine down to one, just the odd numbers, and just the even numbers. Nothing. Somehow we needed a clue. After days of helping John move some things around, it finally occurred to me where to begin. The circle could have been embossed by itself, but it wasn't. Whoever embossed the leather actually had to spend extra time making the diameter line. That was a clue. It had to be significant. A circle with a diameter brings to mind that most magical number pi, and I really should have thought about that faster.

Pi is a fascinating number, especially if you stop and actually think about what it does—which most people don't do. It's amazing how uninterested people can be sometimes. I know I said school was a bit boring, or sometimes very boring, but learning itself is exciting and amazing the more you do of it. The problem is that some people stopped being amazed by cool things so long ago that they're numb to amazing stuff.

Before going into more about pi, I feel I have to pause for a moment to bring up something that is utterly fascinating. If you look out at the world, there are only two shapes that constitute everything we see. Just two. Straight lines and curves. We either perceive a straight line or a curve. I don't want to get into the technicalities of whether something is ever actually straight or not, and what you may be using as a reference point. For now, I want to keep it simple. Look out the window. Look around the house or school. Look at images on a television. It doesn't matter where you focus, everything is made up of straight lines and curves.

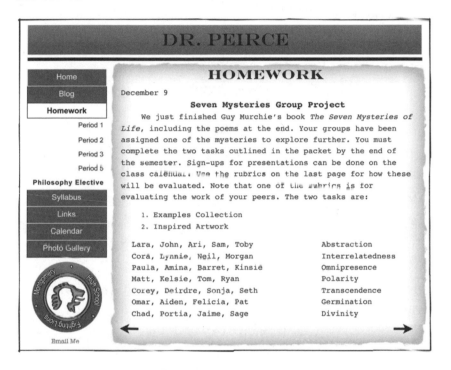

Still keeping it simplified in terms of polarity, nature tends to be made up of curves. Think about meandering streams, drops of dew, oranges and pomegranates, the sun, the moon and many more examples. Think about the spiral forms of shells, the waves in water, or the curves in clouds. And then there are the cycles of time. Think about the phases of the moon or the reoccurring tides. Seasons come and go only to come back again. I know there are straight lines in nature, such as sodium crystals, the horizon line on the ocean, or an occasional rock face, but it seems like a majority of things in nature are based on the curve.

Human-made things, on the other hand, tend to be based on straight lines. Think about landscapers and the sophisticated instruments specifically designed to make things level. Or a carpenter's *square* to create right angles. Somehow an angle is "right" to people if it forms perfectly straight edges and lines. Think about highways, baseball diamonds and parking lines. Time in the human world is linear as well. Even though, the seasons come and go we mark them with a linear calendar year. And, of course, think about the houses we live in which are quite literally boxes.

Pi is the bridge between the polarity of the two worlds of curve and straight. It attempts to connect the world of the straight line (the diameter of a circle) with the world of the curve (its circumference). When I think of pi this way, I think about blending the two worlds together to form infinite possibility for humans. Think about yin and yang and the balancing of those natures. Think about straight line as masculine and curve as feminine. Our world needs both obviously. I say *attempts* because pi is a transcendental number, basically meaning that it is infinitely long. In this sense, pi isn't "normal" because it combines both worlds. The infinity symbol along the spine suddenly made sense because pi is infinitely long.

Looking at the journal, I punched in the typical numbers for pi: 3, 1, 4, 1, 5, 9, 2, and 6. Nothing. I tried pi out farther and farther until I reached the 32nd digit and realized there was no zero button on the cover. Still nothing. Something began to bother me about the infinity symbol.

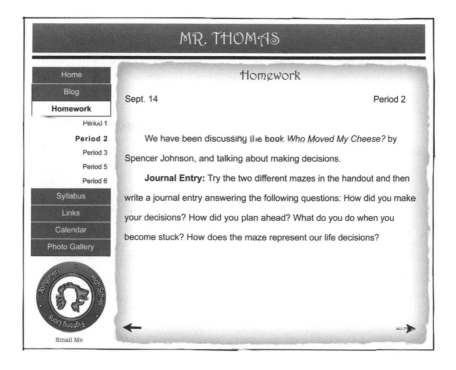

In math, when things don't work out like I just experienced, you have to backtrack and re-approach. I think about it like landing an airplane. If the angle is wrong, then the plane has to pull up, circle around and try again. The same holds true of working out math problems. You can get into a line of thinking that influences the way you work with a problem. If the approach is wrong, then at some point the line of thinking will dead end. A lot of people make a classic mistake at this point. They will try to stay in the cul-de-sac looking for a way out, as if the world will re-shift like a brick wall from Harry Potter. In a testing situation this can waste a crazy amount of time, as many of my friends have discovered.

This happens all the time to people when they get stuck, and I don't just mean in mathematics. But let's take a math problem. When you work a math problem, you start with an initial idea and add steps. Even if you have several ideas in the beginning, you usually have to pick one in order to proceed. Each step that you add might seem perfectly logical to you, otherwise you probably wouldn't have done it that way. But if you make a mistake somewhere, then you stray from the path of solving the problem, only you don't know it.

This is why math teachers always say the famous slogan, "Double check your work." It doesn't mean test your answer using the exact same thinking that you first used. That, of course, would likely produce the exact same *wrong* answer. What they mean is *double* the ways that you answer the problem, so that you can check your own *thinking*. In a way you're not checking the math problem, or even the answer. What you're really doing is checking your own thought process. Here's the issue: you have to think about your thinking in order to do this. Most people I know are basically too lazy. If they do double-check their work, what they usually do is lazily gloss over what they just did and follow the maze right back to the same dead end again.

The problem is that they need a new and different *line of thinking*, only they either don't know they've reached a dead end, they never realized they may have missed some twists and turns before, or maybe they just stubbornly keep thinking that they can discover a secret door in a blank wall or something. Okay, maybe that's not a good analogy considering my circumstances, but my point is that in my situation, I realized I was at a dead end and I needed to backtrack and "double check" my work.

So, I looked at the cover again. A circle with a diameter. Two semi-circles. Two halves. I suddenly had a new approach, a new line of thinking. A circle is composed of 360 degrees, so half a circle is 180. That must be it. I excitedly reached out to press the digits 1-8-0 and then realized there wasn't a zero button. Luckily, I hadn't pressed anything.

I needed another new approach. What if the circle represented a clock face? Then the diameter could represent the hands of the clock, one at twelve o'clock and the other at six o'clock. That would mean pressing 1-2-6, or maybe 6-1-2. I tried 1-2-6 thinking it might work.

"Matt!" John was shouting at me from the corridor. I looked up. "Man, I've been calling you! We have to go, it's already six."

I could hardly believe it was six o'clock. I'd been thinking about this journal for almost four hours. I thought about taking the journal with me and then decided against it. We had discussed taking things out of the lab, but John had convinced us that taking things increased our chances of getting caught. A lot of his reasoning included references to movies where there were locator beacons and spy devices in everything. I thought he was a little too paranoid, but he convinced us in the end. Besides, I didn't actually need the journal to work out the math.

I took a last look at cover of the journal. Did I need yet another new approach, or had I missed something in one of my attempts?

I guess it's no surprise that I couldn't sleep that night. I went through everything I could think about circles and diameters: astrology symbols, alchemical symbols, clocks, and more. Somehow nothing seemed right. I decided to backtrack all the way to my original idea that the cover was supposed to represent pi. I thought about it in the context of the journal itself. No matter where I went with my thinking, I kept coming back to pi.

Finally, I knew what I was missing.

Pi is normally represented as 3.1415926, or 3.14 rounded to two places, which is fine for most people's purposes. But that is only one version or interpretation. Pi has other approximations that can be used, particularly when working with fractions. As an irregular fraction, pi can be approximated as 22/7, yielding the decimal 3.142857. Numerically this is actually more interesting than the standard formulation of pi, mainly because of division by the number seven.

Seven has a lot written about it. There are seven days in a week, seven colors in the rainbow, and seven notes in the Western major music scale, which I find fascinating because I can't help but think that each day of the week has a color and a musical note that goes with it.

There are the seven heavens, the seven sins and the seven sacraments. There are the seven seas, the seven continents, and the Seven Wonders of the World. There are seven energy chakras, seven openings in the heads of most mammals, and seven classification systems in scientific taxonomy. Everywhere you look, you can find examples of the number seven. Collect them in a notebook and you'll not only begin to see the number seven everywhere, but you'll also make connections between things you never thought about before. It's actually a lot of fun.

I made an app for my iPad to collect and organize examples. Here are a few:

1. Seven vital organs in the body.
2. Seven days of the week.
3. "Seven year itch" and seventh inning stretch.
4. Seven liberal arts.
5. Rome was built on seven hills.
6. Seven Years War and the Seven Days War.
7. Seven main types of dogs.
8. A menorah has seven branches.
9. Achilles dragged Hector's body seven times around Troy.
10. Seven major kinds of lightning.
11. Seven voyages of Sinbad.
12. $1 \times 2 \times 3 \times 4 \times 5 \times 6 \times 7 = 7 \times 8 \times 9 \times 10$.
13. Seven stations of the cross.
14. "One, two, three, four, five, six, seven, all good children go to heaven."
15. Buddha meditated seven years under the Bo tree.
16. The Magnificent Seven.
17. Carbon has seven major structural types.
18. Cleopatra VII was the famous one.
19. Seven deadly sins.
20. Seven parts to a dictionary entry.
21. There are seven rows in the periodic table of the elements.

One of the crazy properties of seven is that numbers (except zero and multiples of seven) divided by seven form the same repeating six digits: 142857. Take the number one and divide it by seven. Answer: .142857142857 ... Now try 2 divided by seven. It's the same six digits, but in a different order: .285714285714 ... The same is true with dividing three, four, five, six, eight or nine. They are the same six digits in different order. These are called cyclical numbers because they repeat themselves in a cycle—in this case six digits repeating. Notice that the cycles are the same for 1 & 8 and 2 & 9, which leads to another pattern ...

Number	Quotient	Cycle
1 ÷ 7	.142857142857142857...	142857
2 ÷ 7	.285714285714285714...	285714
3 ÷ 7	.428571428571428571...	428571
4 ÷ 7	.571428571428571428...	571428
5 ÷ 7	.714285714285714285...	714285
6 ÷ 7	.857142857142857142...	857142
8 ÷ 7	1.142857142857142857...	142857
9 ÷ 7	1.285714285714285714...	285714

By the way, an interesting fact about seven is that it is the only number from one to ten that cannot evenly divide the degrees of a circle. Circles have three hundred sixty degrees in them, which is mysteriously close to the number of days in a year. Dividing three hundred sixty degrees by one, of course, yields three hundred sixty degrees. Dividing by two yields one hundred eighty degrees. By three: one hundred twenty degrees. By four: ninety degrees. By five: seventy-two degrees. By six: sixty degrees. By eight: forty-five degrees. By nine: forty degrees. And by ten: thirty-six degrees. Dividing a circle by seven, though, yields a quotient of 51.428571428571 degrees—and there are those six digits repeating again. A quick digression: a lot of people have pointed out that this angle of 51.42 degrees is remarkably close to the slope of the Great Pyramid of Egypt, which is approximately 51.85 degrees.

With this thought process, I understood that the symbol along the spine was referring to an infinite cycle. In this case, the six digits from 22÷7, which yields 3.142857142857. I texted Ari a message to meet me before school tomorrow and then I was finally able to fall asleep dreaming of all of the ninja things that we might find in this book-computer thing.

I went in to school early and met up with Ari. "You figure it out?" was all he said.

"Good morning to you, too."

"Sorry," he muttered, "I've been a bit distracted ..."

I was about to exclaim, "I know what you mean!" in an excited tone. I was thinking that he meant he couldn't sleep either with all the cool stuff in the lab. Then I looked at his face and knew he was talking about his parents. They must be fighting again. "You want to talk about it?"

"Nah, let's just get this friggin' book open. Come on." He looked down the hall and then slipped into the bathroom. I followed.

Once we were in the lab, Ari walked past the book and shuffled over to the stereo. I went straight over to the book, but waited for him as he picked an album and inserted into what we started calling the upturn table. An interesting song came on that mixed rock music with, of all instruments, a flute. The combination was weird at first, but grew on me. I looked at Ari. "Jethro Tull," he said as he walked over.

"Never heard of him," I said as I repositioned the book so it was oriented in between the two of us.

"*Them.*"

"What?"

"*Them.* It's the name of the whole band."

"Oh," was all I managed.

Ari shifted in place. "Okay, Matt, let's see your magic."

I thought about the cycles from last night. Since the whole number 3 is not part of the infinite series, I skipped it and pressed the buttons for 1-4-2-8-5-7. There was a humming noise and then all of the pages were suddenly backlit with the beauty of digital light.

9

Dyna-Reading

Ari and I eagerly flipped open the book. I'm not sure what Ari was expecting, but I was already thinking about all the amazing things that must be in such a cool book. I had already thought about secret government blueprints, CIA cover-up documents regarding Area 51, special NASA records about Mars landings, and many more.

The pages inside were filled with images and patterns that I immediately recognized as mathematical puzzles. Some were charts of numbers; others were made up of pictures and diagrams. At first I was a bit excited, but then I thought about the code on the sink, on the lab door and on this journal. "Let me see those," I said to Ari. He handed me the stylus-calipers, and I poked one of the screens hoping that it would be responsive. Nothing. Just as I was afraid of: we were going to unlock each page by solving a puzzle. The one I had flipped to showed blank puzzle pieces arranged in ten columns and ten rows. In two rows, one above and one below, were nineteen numbers: 3, 5, 11, 13, 19, 23, 29, 37, 41, 47, 53, 59, 61, 71, 73, 79, 83, 89, and 97.

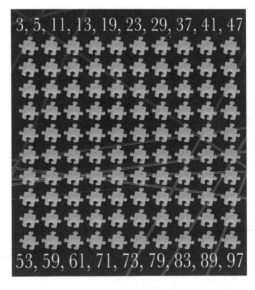

I tried a few more pages, but every screen was locked with various password protections. When Ari saw that the screens were locked, he wasn't happy. "Are you serious? More freakin' codes!"

"At least we got it working."

"Yeah, but we've already spent months getting in here, weeks fiddling with stuff, and days just turning this on."

"Following your pattern, it should only take us hours to crack some of these."

"Oh, be quiet, Matt. You know what I mean." When I didn't say anything, he continued. "I'm with Tom and Jamie, man. We already have a secret lounge in our school to hang out in. You know how many people would kill to have something like this?"

"Thirty-seven thousand, three hundred, thirty-seven." I love answering rhetorical questions. By the way, 37,337 is an interesting number because it is a right-truncatable prime number, meaning that if you stopped writing it at any point, you would still have a prime. Starting from the ten thousands place, the 3 is a prime number. Then 37 is also prime, followed by 373 and then 3,733 and, finally 37,337 itself is prime as well.

"What?"

"You asked how many …"

"Oh, shut up, Matt! I'm done with this Sherlock Holmes crap." He stormed over to the living room area and dove dramatically onto one of the couches. A cloud of dust erupted around him, and I watched as it settled.

I guess that settles it, I thought.

I turned back to the screen and focused on the numbers 1-100 in the 10 by 10 grid. I thought about the nineteen numbers in the top and bottom rows: 3, 5, 11, 13, 19, 23, 29, 37, 41, 47, 53, 59, 61, 71, 73, 79, 83, 89, and 97. Out of curiosity, I used the stylus and poked those numbers on the grid in ascending order. A message flashed briefly across the screen:

ACCESS ... DENIED ... TWO MORE INCORRECT ENTRIES AND DATA WILL BE LEVEL NINE ENCRYPTED ... FOUR INCORRECT ANSWERS AND DATA WILL BE UNINTEGRATED.

"Um, Ari, what do you think unintegrated means?"

"*Unintegrated*?" he called from behind the couch.

"Yeah."

He sat up. "What did you do?"

"I just tried something."

Ari's head was bouncing to the music of Jethro Tull. "Unintegrated, huh? Be careful, Matt, I'm guessing all the data in that thing would be completely destroyed."

"You mean deleted?"

"No, worse."

"*Worse*?"

"Well, computers don't usually erase information completely. They just use the same space to store new information. I've heard good hackers can often recover deleted files because they're still kickin' around as bits and bytes."

"So …"

"So, *unintegrated* probably means *completely destroyed* somehow. What did it say, exactly?"

I repeated what I had seen: "ACCESS … DENIED … TWO MORE INCORRECT ENTRIES AND DATA WILL BE LEVEL NINE ENCRYPTED … FOUR INCORRECT ANSWERS AND DATA WILL BE UNINTEGRATED."

"Oh man! Be careful, Matt! Level nine encryption sounds bad enough. These puzzles have been super annoying. Don't make this any worse."

I was going to say something bitterly sarcastic, but held my tongue. I have found that sarcasm rarely helps anything. "Don't worry," I said, "I can get this one."

As it turned out, I was able to solve the first one the next day at school.

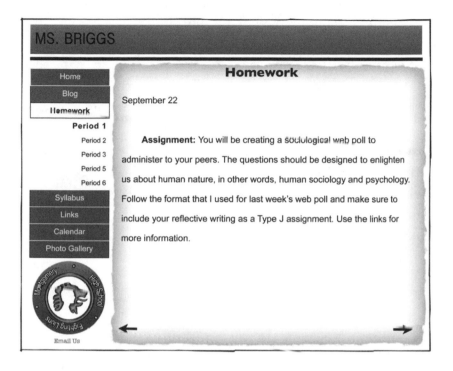

I thought about the nineteen numbers around the perimeter, and they seemed random at first. The problem is that a lot of seemingly random numbers only seem random because there is no known pattern to the observer holding them together. The trick is to find a pattern to explain the apparent "randomness." I know there are random number generators on computers and you can, of course, do something simple like rolling dice. But even with those two examples there is a pattern from the origin of the numbers. For example, a computer has to be programmed in order to generate random numbers, so it's possible to examine the software and reverse engineer *how* they are being generated.

The other problem is that we generally have a sense of what is *supposed* to be random. In the case of rolling dice, for example, let's say you rolled a six-sided die. The numbers generated would be random in the sense that the die tumbles around and lands on different sides in a seemingly random way, but it would always roll a number from 1-6. It couldn't, for example, roll a 97. There is also the problem of whether the die is equally "random" or not. Small defects in the edges or unequal weight distribution can cause specific numbers to appear more than others, disturbing the probability distribution.

If we can find a definite pattern, then we usually rule out randomness, but if we can't find a pattern, then we *might* be willing to accept them as random. Here are some examples with rolling a single six-sided die. If I told you I rolled the die twelve times and came up with the following patterns, you probably wouldn't believe that it was actually a random sequence:

A_1: 1 — 2 — 3 — 4 — 5 — 6 — 1 — 2 — 3 — 4 — 5 — 6; or

B_1: 1 — 1 — 1 — 1 — 1— 1 — 1 — 1 — 1 — 1 — 1 — 1

These two sequences are held together by too much of a pattern, very noticeable at that. If the patterns were broken up just once, most of the students I polled still wouldn't buy into the idea that these were randomly generated.

A_2: 1 — 2 — 4 — 4 — 5 — 6 — 1 — 2 — 3 — 4 — 5 — 6

B_2: 1 — 1 — 1 — 1 — 1— 1 — 1 — 1 — 1 — 2 — 1 — 1

How about one more change? When I polled most of my classmates, 92.59 percent answered that they still didn't believe these were random enough.

A_3: 1 — 2 — 4 — 4 — 5 — 6 — 1 — 2 — 3 — 5 — 5 — 6

B_3: 1 — 1 — 1 — 6 — 1— 1 — 1 — 1 — 1 — 2 — 1 — 1

After a few more changes, only 26 percent of my classmates started believing they could have been randomly created.

A$_4$: 1 — 6 — 4 — 4 — 5 — 2 — 1 — 2 — 3 — 5 — 5 — 3

B$_4$: 1 — 4 — 1 — 6 — 2 — 1 — 3 — 1 — 4 — 2 — 1 — 5

The problem, of course, is that any of these sequences could have been randomly generated. People are just more comfortable with certain sequences being random and are more skeptical of others.

Anyway, I digress. The point is that the numbers on the page bothered me because they *seemed* random, but I knew they weren't. Somehow, they formed a clue.

Counting carefully, there were nineteen numbers in the two rows, and nineteen has interesting significance in various places. Here are some that I collected in my numbers collection:

1. The 19th Amendment to the Constitution gave women the right to vote.
2. In the Qur'an, there are 19 angels that guard hell.
3. In the Baha'i Faith, the calendar is divided into 19 months with 19 days each.
4. There is a symbol called the flower of life that has 19 interlocking circles.
5. A 19 × 19 board is standard in the game of Go.
6. A 19-gun salute is used for deputy heads of state.
7. A "19" in cribbage is slang for a worthless hand.
8. $19 = 4! - 3! + 2! - 1!$
9. The 19th hole in golf refers to the clubhouse bar.
10. The periodic table number for potassium.
11. 19 is the number of the Sun card in a tarot deck.
12. Eclipses of the sun seem to occur in periods of 19 years.
13. My dad's favorite hockey player, Rod Brind'Amour, wore #19.
14. The number 19 is a twin prime with the number 17.
15. In Bingo, ball 19 is called "Goodbye teens."

16 In Roman numerals: XIX.

17 In the Bible, Matthew 19:19 "Honor your father and mother."

18 The 19th president of the U.S. was Rutherford B. Hayes.

19 August 19, 1919: Afghanistan gained independence from the U.K.

For a lot of kids, math seems like just a tedious repetition of basic arithmetic. But math is a much bigger realm than people suppose and many aspects are much more interesting. Basically, math is about finding patterns. It's about persisting through confusion until the answer is so clear that it must be right. Take the Fibonacci sequence, for example:

0, 1, 1, 2, 3, 5, 8, 13, 21, 34, 55, 89, 144 … and so on.

It's a pattern. If you don't see it yet, that's okay. It took me a while to develop a knack for seeing how things relate to each other. See how in the beginning the numbers grow slowly and then they start to grow faster? Then observe even more. See how much bigger the numbers are getting each time? Now notice that the number 1 is repeated. That's the big clue in this case. It's the piece that looks odd. It's the part of the pattern that doesn't seem like it would hold true. So, in terms of the pattern, we want to find something that is always true of the whole problem, not just an isolated part.

With the Fibonacci Sequence it's that each number after 1 is created by adding the previous two numbers: 0 + 1 = 1, 1+1 = 2, 1+ 2 = 3, 2 + 3 = 5, 3 + 5 = 8, and so on. So I looked at the number 19 not in isolation, but in relation with the other numbers along the spine of the journal. That's when it dawned on me that all of the numbers were prime.

Besides Cardinal & Ordinal and Even & Odd, there is another binary way of thinking about numbers: Composite & Prime. Primes are numbers that only have factors or divisors of one and themselves. So, seven is a prime number because it can only be divided by one and seven without a remainder. Composite numbers are numbers that have more than two factors or divisors. Take eighteen, for example, it has factors of 1 and 18, 2 and 9, 3 and 6, so it is a composite number. Prime numbers are *prime* because multiplying prime numbers can form the composite numbers! This is a process called Prime Factorization.

This is how it works. Take any composite number and pull out the factors that are prime numbers. Repeat until all that are left are primes. For example, with eighteen, there are factors of 3 and 6. Three is a prime number, but six isn't because it has factors of 2 and 3. Since those are both prime numbers, the prime factorization is done. So, 18 = 3 × 2 × 3, or in shortened form: $3^2 × 2$.

Let's take a larger example, like seventy-two. 72 is 9 × 8, but those are both composite numbers. 9 = 3 × 3 and 8 = 2 × 4 and then 4 = 2 × 2. So the number 72 has prime factors of: 3 × 3 × 2 × 2 × 2, or $3^2 × 2^3$ for short. All primes. The process is a little easier to see as something called a factor tree.

$$72$$

$$9 \quad × \quad 8$$

$$/\backslash \qquad\qquad /\backslash$$

$$3 \quad × \quad 3 \qquad 2 \quad × \quad 4$$

$$/\backslash$$

$$2 \quad × \quad 2$$

Okay, back to the journal. There were two issues. Two is a prime number, but Ari noticed it wasn't included in the list. Seven wasn't listed either. Those missing primes were the next clue. The series couldn't simply be the prime numbers because two of the first few were missing. The grid had one hundred puzzle piece and from 1 to 100, there are 25 primes. Out of those, there are 25 primes. By the way, you can figure out prime numbers using something called the Sieve of Eratosthenes, which was developed by an ancient Greek mathematician. So I looked at the list of nineteen numbers around the perimeter. Six were missing. $25 - 6 = 19$. So the missing prime numbers were: 2, 7, 17, 31, 43, and 67.

Now the question was: In what order should I press the buttons? Well, I wish I could say that I spent more time thinking about that, but you've already gathered that I'm a bit impulsive. I pressed the buttons in ascending order and half expected a trap or something, but the screen simply unlocked. No one else was in the lab with me at the time, so the whole thing was a little anti-climatic.

I grabbed the stylus and held it like chopsticks as the screen re-freshed. I found myself staring at twenty little pictures of books arranged in a grid of four columns and five rows. I used the stylus to poke one of the images. The book opened and then enlarged until its digital pages filled the screen. I was now looking at a table of contents with a huge list of dates. I touched one of the dates with the stylus and a text document sprang open. I scanned through some of it, but it appeared to be a boring journal entry about someone named Maglio who started working for a company called the Aerospace Research and Education Association on May 2nd, 1992. I laughed to myself as I realized the acronym for the company, AREA, and the date 5/2/99, could sort of be combined into Area 52.

I navigated back to the initial grid of book icons and selected a few more, but they were all journal entries by Maglio. I wasn't very interested in the stuff because most of it was just junk about how mistreated this guy was, how his ideas were never taken seriously, and all this other boring stuff. I was really hoping for something far more interesting, but at least I had the other screens to still unlock.

I had one of the journal entries open when I wondered why the stylus was two-pronged. I carefully placed both ends on the screen and pinched the ends together. The text on the screen instantly scaled down to 65 percent. I released the ends a bit and they sprang apart, scaling the document to 87 percent. I lifted them off the screen, pinched them together completely and then touched them again to the screen. I let them spring open completely, and the document scaled to 233%. Pretty cool!

I had a nagging suspicion, though, that it would be more functional than just scaling things. I thought about mathematical compasses and how they are used to draw circles by jamming one end in the paper and spinning the other pencil end. I took a look at the two-pronged stylus and thought it would be worth a shot. I put one end in the middle of the journal entry on the screen and then carefully placed the other end down. Then I spun them clockwise. I wasn't at all ready for what happened.

The moment I spun them, I felt a shock race up through my hand, through my arm and into my chest. The words lifted off the screen and hovered three-dimensionally in the air. Some of them were just barely off the screen, maybe five millimeters or so, while others rose up three, even four centimeters. They shimmered and took on various colors and intensities. What had simply been black text on a flat screen became a colorful landscape of words and sentences. There was a series of red adjectives that formed a mountain range, and a group of blue scientific words that formed a valley. There were offshoots of yellow and green pools and flat expanses. I was so shocked that I dropped the stylus device and the words "fell" back onto the screen.

After the initial shock wore off and I convinced myself I wasn't hallucinating, I cautiously picked it back up and repeated the spinning, only this time I held on long enough to spin it a full three hundred and sixty degrees. The words popped off the screen again and returned to the same 3-D landscape. I was focusing on the tallest word, *frustrated*, as I completed the revolution, and that's when this amazing experience became even cooler—or creepier, depending on the viewpoint.

What happened was that a cascade of emotions hit me, nearly knocking the wind out of my lungs. I was choking for air as waves of thoughts and feelings stunned me. I felt the entire emotional state the author experienced at the time of his entry: I could feel the rapid pulse of his heart, his shallow breaths, his clammy sweat. I could feel all of his historical frustrations as they developed into the current writing, could sense the evolution of his powerlessness, and, most of all, I felt his building anger. In just a few seconds, I had virtually experienced months of his life.

After I got accustomed to the emotional content, I focused more on the individual words. Somehow they were each encoded with information. I knew, for example, when a word was sarcastic or serious, and to what degree. I realized the taller words had more history with the author and that the shades of color indicated levels of power or intensity. I could tell which words were his favorites, which ones he had trouble spelling, and which ones he thought were funny or strange. In mere moments, I had a deep and profound understanding of this one journal entry page, probably equivalent to an entire novel or something.

I scrolled through and "experienced" a few more pages. Then I went back to the main menu grid and opened a few of the other book icons. All the journal entries had similar feelings of anger, frustration, and isolation. Despite how cool this new experience was, after only twenty minutes, I was practically exhausted. It was as if I had read a dozen novels about Maglio's life without stopping, or more accurately, maybe I had truly gone through what Maglio had. In any event, it was a pretty gross experience, so I was happy to go home and take a shower.

On the way home, I texted the others: *1st screen done meet b4 school.*

10

Triangles and Boxes

Choosing the next puzzle to work on was easy, since Ari discovered something really cool about the last few pages. Rather than simply flipping open along the binding, like the others, the last few pages unfolded to form a box without a lid. They even magnetically snapped together to stay fairly rigid. We were all fascinated thinking about what the box would be capable of doing, especially considering how cool the stylus alone was.

So, the next day after I showed the others the cool stylus thing and we had a great time making the text "jump and jive" as Thomas put it, Jamie and I snuck into the lab during study hall. We took a look at the puzzle on the last page, and then he shrugged his shoulders and went over to play some albums. "What are you in the mood for, Matt— Queen or Black Sabbath?"

"Um, Queen I guess." I wasn't familiar with either, but Queen somehow sounded more appropriate.

Jamie put on a record and then flopped onto the couch.

"Hey, aren't you going to help?"

"It's in my head, don't worry," he said. But I figured that meant he was going to fall asleep on the couch.

The puzzle was repeated on the five unfolding pages and consisted of a series of size-increasing equilateral triangles. A simple keypad with the numbers 0-9 was underneath. Inside the largest triangle was T57. That was it.

I thought at first that maybe this was a tangram or rearrangement puzzle of some kind, so I tried to use the calipers to move them around, but they couldn't be manipulated.

T57

I thought about the triangular form itself. Triangles are extremely important for a lot of reasons. They are structurally very strong, which is why they appear so many times in architecture. Just look at huge electricity towers or bridges and notice the intricate patterns of triangles. The reason is because triangles can't be deformed as easily as other shapes. Squares, for example, can somewhat easily be bent at the corners to deform into rhombuses. For example, take a cheap picture frame without the glass, wiggle it a bit and notice that the corners will begin to loosen. Add a little force, and the frame will distort into a rhombus, which is exactly what we don't want to happen with things like bridges.

Triangles, though, are more stable. With a square, the distortion occurs with the *angles*, not the sides. The pieces of wood in the picture frame, for example, do not have to lengthen or shorten in order to become a rhombus. But with a triangle, the sides would also have to change—meaning they would have to break. So, in this simplistic way of thinking, a square's strength is only in its angles, whereas a triangle has strength in its angles *and* sides.

Triangles are also important because there's a whole branch of mathematics, called trigonometry, dedicated to the study of them. It seems hard to believe, but it's true. The main functions derived from triangles, namely sine, cosine and tangent, are used in all kinds things: music theory, acoustics, optics, probability, statistics, seismology, meteorology, phonetics, economics, cartography, game development and, of course, engineering. There are also really cool applications in space exploration with something called spherical trigonometry. What's completely beast is that you can draw a triangle with three 90-degree angles on a sphere—meaning a triangle can have 270 degrees!

Then my mind jumped to thinking about famous triangles. I know that sounds weird, but there actually are famous triangles. There's the Bermuda Triangle, which is the most famous. There's another similar triangle off the coast of Japan called Dragon's Triangle and another one over the Great Lakes called the Great Lakes Triangle. Then there are two well-known mathematical triangles, the Sierpinski Triangle and Pascal's Triangle.

Clearly, there was some kind of pattern with these triangles that could be changed into numbers to press on the keypad. There were three of the smallest triangles, two of the next size and one of each of the next sizes. That produces the series: 3, 2, 1, 1, 1, 1, 1, 1, 1 … I doubted it was the right way of thinking about this problem. I pressed a three, a two, and a bunch of ones into the keypad anyway, but it didn't unlock.

I thought it was interesting that after the first few, the triangles increased in size just enough to create a border for the previous two triangles. Somehow, I felt like this was a pattern like the Fibonacci

Sequence or Pascal's Triangle, but I couldn't figure it out just by looking at the images. Somehow I was missing how to convert them into a numeric pattern.

A new song came on the cool record player and then I thought about the Queen album. Ruler. I grabbed a ruler out of my backpack and then measured each of the triangles. They were all based on the meter, or more precisely, centimeters. The series went: 1 cm, 1 cm, 1 cm, 2 cm, 2 cm, 3 cm, 4 cm, 5 cm, 7 cm, 9 cm, 12 cm and 16 cm.

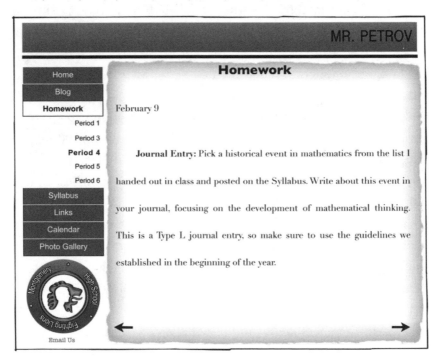

The meter, or metre as it is spelled elsewhere in the world, has an interesting history, mainly because of its definition and, therefore, over time, its length has changed several times. I never imagined when I first picked up a meter stick that there was such a long history. There were two early definitions of the meter, one based on a swinging pendulum and the one that I learned in sixth grade, which defined the meter as one ten-millionth the length of Earth's meridian from the

Equator to the North Pole. In 1793, the French government even commissioned an entire expedition to travel the meridian in order to make a standardized length. Apparently, they made a miscalculation and created a meter that was .2 mm too short.

Eventually, an even more specific standard was defined as the length of the path light travels in a vacuum in 1/299792458 of a second. In other words (or numbers), light was defined as traveling 299,792,458 meters per second.

It might not seem like a big deal—a little deal, I suppose, is more appropriate—but even if two definitions of a "meter" varied by .0001, there could be disastrous consequences. Imagine if there were two teams of engineers working on the world's longest bridge, the Danyang-Kunshan Grand Bridge in China. One team working on the left side of the bridge uses a standardized meter, and the other team accidentally uses an older definition that is .0001 too short. The bridge is an amazing 164,800 meters long, so such a difference would be $164,800 \times .0001$, or 16.48 meters too short! The math makes the point: tiny differences can build into large discrepancies.

Anyway, with the measurements of the various triangles in the pattern, the sequence of numbers went: 1-1-1-2-2-3-4-5-7-9-12-16. I had been thinking about Pascal's Triangle and the Fibonacci Sequence, so I did finally recognize a pattern. The three ones in the beginning gave me a problem until I realized that they were just the starting set up for the sequence.

I suddenly noticed that the music wasn't playing. Jamie called out, "Yo, Matt, study hall's almost over. We have to get to class."

I looked up from the journal. I figured the T57 was the 57th term of the sequence, so I needed to program a formula. "But I just need to program my calculator and then …"

"We have that big quiz today. Just finish early, and then program it."

I grabbed my backpack. "Yeah. I found a pattern, but I think we have to find the 57th term."

"Fifty-seventh? Why 57?"

I shrugged. "Maybe it's important."

"You mean besides activating these pages?"

"Maybe," I said as I entered the narrow hallway back to the boys' bathroom.

"What's the pattern?"

"I thought it was in your head."

He shrugged, "*Was.*"

We checked the bathroom and when the coast was clear, we snuck out into the hall just as the bell rang. I explained the Fibonacci Sequence again to him as we went to class. He got pretty excited about the whole thing and boasted, "I'm going to make my own pattern for you to solve: Jamie's Sequence."

I laughed, "Okay, but you're going to get Matt's Sequence then."

After our quiz, we both had time to work on the sequences. Sequences are much easier to make than to decode. Generating the numbers becomes a simple matter of plugging them into a formula. Reverse engineering the pattern of numbers, though, is much more difficult because it requires deducing a formula out of the sequence, order out of chaos. I made a chart on one of my phone apps to compare the sequences. Notice, that the unknown sequence builds more slowly than the Fibonacci Sequence, and a lot more slowly than Jamie's.

Sequence	Start	Steps of the Sequence								
Fibonacci	0, 1	1	2	3	5	8	13	21	34	55
Unknown	1, 1, 1	2	2	3	4	5	7	9	12	16
Jamie's	1, 2	6	16	44	120	328	896	1552	4896	12896
Matt's	1, 2	3	5	8	4	3	7	1	8	9

Seeing the different sequences together provided me with another clue about the triangles. In the Fibonacci Sequence, the next number is created by adding the previous two numbers. Jamie's Sequence grew too quickly to be an additive process, so I suspected it incorporated multiplication somehow. Jamie's second term (16) was 2.67 times bigger than the first term (6). The third term was 2.75 times bigger and the fifth was 2.72 times bigger than the fourth. These were close, but not consistent, so I looked for another part of the pattern. I focused on the starting component and $1 + 2 = 3$ and three doubled was 6. So there was an addition and a multiplication. The next term was $2 + 6 = 8$ and eight doubled was 16. Then $6 + 16 = 22$ and twenty-two doubled was 44. The pattern held.

The unknown sequence didn't grow very fast, so I suspected it only employed addition. It might have been something complicated, such as multiplication by one number and division by another, but I figured I would start with simpler explanations first. The next clue was that the unknown sequence started with three numbers and had a repeat answer in the first and second terms. Then I realized that the pattern was remarkably similar to the Fibonacci Sequence, except that each new term was the sum of the numbers from two and three terms ago.

I finished creating my formula, plugged in $N = 57$ and got: 4983378. "I think I have the code."

"Already, man? The period's not even over and I'm still working out your sequence." Jamie looked up with a pained expression on his face, "Can I get a hint?"

I laughed, "Sure. Notice that all of the terms are a single digit."

"I know *that*. I thought I had a pattern. One and two makes three, two and three makes five, three and five makes eight. But then five and eight makes 13, not 4. That's where I'm stuck."

"So, how are 13 and four related?"

"What?"

"How are they related?"

"That's what I'm asking here, Matt."

Just then the bell rang. "Come on," I said, "we have a 3-D journal to unlock."

We had to wait until after school to get back into the lab, and somehow Kelsie was already there waiting for us. "About time," she said getting up from the couch. "Did you figure out the code?"

"I think so," I said.

"I Googled the triangles thing, and it's something called the Padovan Sequence," she said. "There was a Padovan Sequence calculator link on Wikipedia. I just put in 57 and got 4983378."

"Yeah! That's what I got."

"Well, let's get this party started," she said.

Jamie was over by the *upturn table* and all the albums. "Here's a perfect band for our situation: The Doors." He pulled the album out of the sleeve and then started laughing, "And the first song is called *Break on Through to the Other Side!*"

After a good laugh, we opened the journal, unfolded the last pages and magnetically clipped them together to form the box. The triangle puzzle with the keypad was on all four of the side screens and the bottom. I was about to press the buttons on the keypad, but Kelsie grabbed my wrist. "Wait, Matt, can I have the honors?"

"Sure, go ahead."

She paused, "What do you think is going to happen?"

I was about to answer that I had no idea, but Jamie chuckled, "It's probably going to Rickroll us or something."

Kelsie ignored him and then leaned in and pressed the numbers: 4983378 on the keypad. The screens went blank and I thought we did something wrong, or tripped some alarm, but then a three-dimensional holographic image of some kind of airplane appeared in the middle of the box. The outer screens of the box filled with numbers and charts, which caught my attention immediately. I caught a reference to *coefficient of drag*, so I suspected that all of the data had to do with the craft's aerodynamics.

Jamie reached his hand into the box and poked the hologram with his finger. I didn't know what to expect, but disappointingly, his finger just passed through it. The hologram rippled, and some of the data changed, but then everything returned to how it was when we first saw it.

"Is that it?" Jamie whined.

"Well," Kelsie muttered, "it's pretty cool—isn't it?"

I was about to remind them how amazing this thing actually was—far cooler than our spotted composition notebooks, for example. After all, it was a journal with digital pages that could be folded into a box to make holograms! Then I had an idea. "Wait," I said, reaching for the calipers, "what do these do?"

I reached into the box and poked the hologram. Instead of passing through, the calipers "stuck" to the image and dragged part of the aircraft's wing out farther. The data charts went crazy and a few of the numbers turned green, but most shifted to orange and bright red. The wing on the other side remained the same, so the new image was horribly lopsided.

"Cool!" Jamie exclaimed. "Make the other wing balance out."

I touched the other wing and tried to pull it out in the same way, but I went off on the wrong angle and the wing became longer and narrower. The data charts shifted again, and a couple of columns changed to green, but most remained orange or red.

"Matt!" Jamie squealed.

"Sorry," I said. "Hold on." I went back to the other wing and tried to make it symmetrical, but I dragged the image too far, distorting the wing past the other one. The charts went almost completely red.

"Matt, at this rate, that thing'll have spaghetti wings."

"You try it, then."

"I could do better with my eyes closed."

"Let's *see* it then."

I held out the calipers, but Kelsie intercepted them. "Wait! I have an idea," she said. "These are calipers, right? Well, let's try something."

She rolled up her sleeve and then propped her elbow on her knee for stability. Then she slowly lowered the calipers down to the wing and gently poked the edge. She then opened the calipers and used them to measure out a more exact length from the fuselage. Then, she repeated her procedure and measured out the length of the other wing. Most of the data streams turned green and when she repeated the process for width, they were all green.

"Surgeon's hands," she said smiling. Kelsie wants to be an exotic animal veterinarian when she finishes school. I suddenly pictured her operating on a lemur and had to stifle a laugh.

"Now, let's see what else we can do," she said as she positioned both ends of the calipers into the middle of the hologram. She opened them up and the fuselage scaled in diameter accordingly. Several columns of numbers dropped sharply and turned red. We sat watching her for a while as she changed just about every feature of the aircraft. When she was finally done, all of the columns were green.

"So ... that means it would fly, right?" Jamie asked.

"I assume so," I shrugged.

Jamie reached out, "What do these do?" He pressed a button along one side of the box. The first button was apparently an UNDO or ERASE button, because most of Kelsie's careful work disappeared.

"Hey," she cried, jabbing the air in front of Jamie with the calipers and snapped them shut dramatically.

"Sorry," he said insincerely, "just trying stuff." He hit another button and the hologram returned to its original dimensions. He hit another button and the hologram of the aircraft was suddenly replaced by a geodesic dome. The columns of data shifted and new charts and graphs appeared on the side screens.

"Oooh, dome ..." Jamie said imitating Homer Simpson's voice. He pressed the button again and now the hologram was a space station or satellite. New charts and grids appeared and I caught something about an oxygen-carbon dioxide ratio. Then there was a submarine, a cannon, a weird windmill thing, an aircraft carrier and something I assumed was a hover tank.

Jamie continued scrolling through the objects when Kelsie suddenly yelled out, "Wait, that looks like the donut!"

11

Joking Around

The next two weeks were a bit of a nightmare. The cohesion we had as a group dwindled away as we bickered and fought about what we thought we were supposed to do. Thomas and Jamie pretty much stopped helping, although once in a while we could guilt them into doing some work. Ari got more and more distracted by family issues, and even when he started sleeping in the lab, he never felt like doing much. John was hyper-focused on rewiring the donut thing and barked at everyone when they didn't help, which meant pretty much all the time with Thomas and Jamie.

We had so many disagreements in The Company that we probably would have argued that one plus one was not two. Well, maybe that's a bad example, because I actually have some ideas about that. When I was at my cousin's wedding ceremony, the priest explained that man and woman would become one. So in that case, one plus one equaled one.

Another way of thinking about this is that one is a straight line and two is either a looping curve or a curve and a straight. In either case, a straight plus a straight do not equal a curve and a straight (or a looping curve). Or you can think about it this way: one and one is three. Let me write it this way: $1 + 1 = 3$. On the left side there are three total symbols, an addition sign and two ones. That's three.

Anyway, the point is that we were arguing about everything. Even Ari and I almost came to blows when we were going to split a two-liter of soda and couldn't decide what to get. I wanted 7-Up, and he wanted Sunkist. As a digression, I have often wondered if there was a series of formulas that were attempted before the final 7-Up. Was there 1-Up? 2-Up? 3-Up? Were those other formulas disgusting or something? Why seven? We decided to get one of those new power drinks instead, but then we got into an argument about whether power drinks are better for you than soda.

Maybe our six wasn't so perfect after all.

I tried to break the tension on many occasions with my arsenal of good math jokes, but I'm not sure they were appreciated. John and I benefited from them at least. I'm not sure what laughing is or does to people, but there is certainly a therapeutic aspect to a really good laugh. Take any really bad situation and find the humor in it somehow, get a really good deep belly laugh and suddenly everything is wonderful.

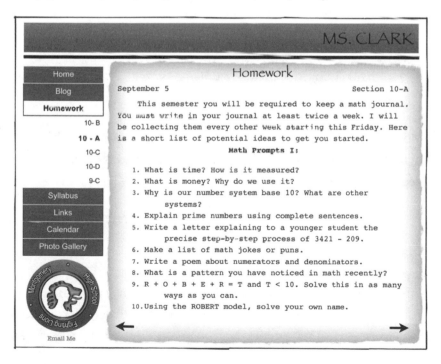

MS. CLARK

Home
Blog
Homework
10-B
10-A
10-C
10-D
9-C
Syllabus
Links
Calendar
Photo Gallery

Email Me

Homework

September 5   Section 10-A

This semester you will be required to keep a math journal. You must write in your journal at least twice a week. I will be collecting them every other week starting this Friday. Here is a short list of potential ideas to get you started.

Math Prompts I:

1. What is time? How is it measured?
2. What is money? Why do we use it?
3. Why is our number system base 10? What are other systems?
4. Explain prime numbers using complete sentences.
5. Write a letter explaining to a younger student the precise step-by-step process of 3421 - 209.
6. Make a list of math jokes or puns.
7. Write a poem about numerators and denominators.
8. What is a pattern you have noticed in math recently?
9. $R + O + B + E + R = T$ and $T < 10$. Solve this in as many ways as you can.
10. Using the ROBERT model, solve your own name.

I know this elevates me to a serious geek level, but I do love a good math joke. I say "elevate," though, because I don't see anything wrong with liking math and math jokes, or, for that matter, being a geek. In fact, I think it makes me special. Not mom kind of special, but special as in different, and different meaning unique.

Here are some of my favorites from the web that I collected on my iPad. Not everyone finds them funny. Kelsie and Thomas have stopped laughing at my math jokes, but Ari and Jamie usually get a chuckle out of the new ones. John thinks most of them are "hi-larious" (as he pronounces it), even when he hears them multiple times. Of course, he still thinks it's funny to say, "That was so punny!" after every joke or pun.

Half of a large intestine = 1 semicolon

1000 aches = 1 megahurtz

Basic unit of laryngitis = 1 hoarsepower

Shortest distance between two jokes = A straight line

1 million microphones = 1 megaphone

4 lawyers = 2 paralegals

2 untruths = 1 paralyze

1 millionth mouthwash = 1 microscope

My absolute favorites, though, are pretty reliable. While the others tend not to laugh at my other jokes or puns, these always make them laugh, or at least smile. I still get a sour look from Kelsie, but I know deep down she loves my sense of humor. Here they are:

Time between slipping on a peel and smacking the pavement = 1 bananosecond

One day, Jesus said to his disciples: "The Kingdom of Heaven is like 3X squared plus 8X minus 9." A man who had just joined the disciples looked very confused and asked Peter: "What, on Earth, does he mean by that?" Peter replied: "Don't worry—it's just another one of his parabolas."

I appreciate people who put bumper stickers on their cars because when you're on a long trip with nothing to do you can at least get a good laugh once in a while from the car ahead of you. One time I did get a good laugh from this one: Math problems? Call 1-800-[(10x)(13i)2]-[sin(xy)/4.3562x]

Around this time, our advanced calculus teacher Mr. Nguyen gave us an evil math problem to solve. He said it's known as the *Three Houses, Three Utilities problem*. He handed us each a sheet of paper with three houses and three utilities. The goal is to draw a line from each utility to each house without the lines crossing. Mr. Nguyen said that anyone who could solve it in at least two different ways would get an iTunes gift card. Sweet! The challenge was on as we all started drawing lines.

Three Houses, Three Utilities Problem
Mr. Nguyen

Goal: connect each house to each utility without the lines crossing.
Hint: There's more than one way!

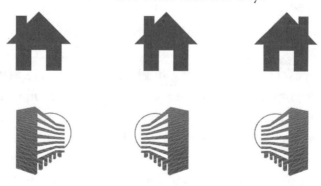

As I quickly realized, there was something diabolical about this problem. I really should have expected it, since Mr. Nguyen loves torturing our little minds with difficult conundrums, riddles and seemingly unsolvable puzzles—like this one. I looked over at Kelsie, but she was busy drawing and erasing lines. I caught Ari's attention, but he just crossed his eyes, stuck out his tongue and went back to work. Then I looked over to Mr. Nguyen and knew we were all in mathematical trouble because he was grinning his evil grin.

I was trying to work out some kind of formula to solve it, but I kept getting stuck. In fact, as far as I could tell the problem was actually impossible. I certainly wouldn't put it past Mr. Nguyen to give us a problem with an impossible solution—especially if he told us it had at least two solutions, and we'd possibly get a gift card. He did things like this to us regularly to "teach us different lessons than our math books."

I was about to "prove" that there was actually no solution to the problem, when Jessica Doolittle raised her hand and annoyingly said, "Can I get that gift card now?" Mr. Nguyen popped out of his seat to look at her work. "Congratulations, Ms. Doolittle," he said, "you are the first to solve this in two years." I never figured her for much of a mathematician, since she was more of the artist type, but somehow she had solved it two ways in less than twelve and half minutes. Amazing, actually. Now, I don't mind getting beat by people, although I'll admit getting whipped in math class stings my pride a bit. But to be beaten by someone named Doolittle seems extra humiliating.

Jessica had two solutions to the problem, and I'm happy to say that after I saw hers, I created a third. I tried to get some credit for mine, but Mr. Nguyen shook his head and handed Jessica the $10 iTunes card. I thought for sure that she was going to waste it on some gross pop songs, but I found out later that she actually purchased some episodes of *Star Trek*. My respect for her beamed up instantly.

Anyway, her first solution involved poking a hole through the paper and sending the utility line down through it, around the other side of the paper and back again. It seemed like cheating, but these were just the thoughts that Mr. Ngyuen was trying to teach us "outside the book." Her next solution was similar to the first. She simply drew the lines off the edge of the paper and back around.

Once I had seen her solutions, I began to think about other ways to think differently. In three dimensions, the companies could simply have the lines go over and under each other. I envisioned the graphite in my pencil coming off the paper and creating little bridges up and over the other lines I had drawn. Then I thought about the houses themselves and realized that there was room in the houses for more than one line. So, my solution was to send a utility line through one of the houses to another house. Problem solved.

By the way, the next day in class, Mr. Nguyen showed us that without using the creative answers Jessica and I had discovered, the problem was actually impossible to solve. He showed us something called the Jordan Curve Theorem and one of Euler's formulas as proof that it cannot be answered on the Euclidean plane. His main point to us was that we had solved it three different ways anyway. *Every problem has multiple solutions, but they don't always exist until we think of them* was one of his favorite mantras. I'm not sure he's totally correct, but I get his point.

Kelsie and I, at least, had a great time figuring out the rest of the puzzles and unlocking the screens. We had a ton of research to do, but we enjoyed the challenges. Most of them involved some kind of clever sideways thinking, but a few involved straight calculations similar to problems in our math books. We had to use calculus to find tangents, velocities and accelerations; we had to use trigonometry to solve problems with angles; we had to use geometry for figuring out spatial questions; and much more.

We spent a lot of time browsing through the various digital pages of the journal. There were holograms, pictures and charts for an amazing array of gadgets and devices like stuff out of a Batman store catalog. There were several varieties of futuristic hover cycles, geodesic houses on hydraulic legs, digital t-shirts, and what we think was a scent-emitting television. There was even a design for a laser-guided toothbrush that would get dentists everywhere drooling in their own spit sinks.

Most of the journal pages, though, were more like diary entries, with long rants and complaints from the author, Maglio. While I spent most of my time unlocking the screens, Kelsie focused on keeping up with all of the text. She thought it was cool that she was reading what could someday be a primary resource from a genius scientist, like reading Einstein's journal before he published anything and became famous.

We were all in the lab one day when Kelsie exhaled loudly and flung the calipers down on one of Jamie's beanbags. "So, Kelsie, did you read all those journal entries?" I asked, putting air quotes around the word read.

"Yeah, just about. But *read* isn't really the right word."

"What do you mean?"

"I don't know. It's just not the same."

"What would you call it?"

"I don't know. Something like emoti-reading."

"*Emoti-reading*? That sounds like what you do when you read stupid text messages."

Kelsie laughed. "How about eReading?"

I shook my head, "Nah, that just sounds like reading off a book reader or something."

"How about empathy reading, sympathi-reading, or something science-fictiony, like Sim-reading or bio-reading."

I shook my head, "Too technical or robotic."

"Hey, did Maglio call it anything?" Jamie called out from over by the albums.

Kelsie thought for a while. "Well, he didn't have a specific name for it, but one time he called it biometric nano-feedback, deep limbic something something …"

116

John yelled over from the living room area, "Dude, just call it mega-reading or super-reading." He sat up from the couch, "No, I've got it: ultra-reading!"

"Been practicing those prefixes, John?" Kelsie asked.

"I'm just saying …" he hollered back.

John disappeared back behind the couch and Kelsie turned back to me. "A prefix isn't a bad idea. What about … dyna-reading?"

"Hmm, I guess," I said. "That's the best we've got so far. We can always …"

Kelsie started giggling.

"What?"

"And if John tries reading, we can call it dyna-*might*!"

"Hey!" he called from behind that couch, "I heard that!"

"So, Kelsie, give us the skinny on the journal entries," Ari said with a mouthful of cheese puffs.

"The *skinny*?"

Ari licked his cheesy fingers, "Yeah, you know, the Cliffs Notes version."

Kelsie rolled her eyes. "If you guys actually did anything around here, it would be amazing."

Thomas and Jamie must have assumed she was talking about them because they both turned. "Hey," Thomas retorted, "we're not stupid. We just fixed up the place so that you and Matt could do all the work." Kelsie looked like she was about to slap him or something.

John interrupted, "What I want to know is: What is that donut thing I've been working on? As far as I can tell, it's the only thing pictured in the journal that's actually here in the lab." Everyone stopped what they were doing and turned to Kelsie.

Kelsie took a deep breath. "Okay, look. I've been doing a lot of, uh, dyna-reading and Maglio spends most of his time on …"

"Cliffs Notes!" Ari yelled.

Kelsie glared at him and then continued, "Most of the entries are more personal in nature. Maglio spends most of his time …"

117

"Ranting and raving, we know," John interrupted. "What about the donut?"

Kelsie exhaled dramatically. "You guys have the attention spans of two-year olds!"

Jamie jumped in, "Wait, did you just say something?"

"Actually, two year-olds …" Thomas started.

I ran out of patience, "Guys, let her finish! She's *trying* to tell us."

"Thank you, Matt," she said kissing the air. "As I was saying: Most of what was written—is written the right word?" Ari was about to say something, so Kelsie quickly continued, "Most of it was about his life: his thoughts, his dreams, his aspirations. Once in a while he elaborates on one of his gadgets, but they're like tangents or something. The interesting thing is that each of his inventions solved some problem in his life. There was this one toaster that …"

"Cliffs Notes!" Ari yelled again.

"Ari!" Kelsie stared at him and made a slashing gesture across her throat. "Okay, the donut thing has a name. Maglio called it a teleportal."

John punched the couch. "Dude, *teleportal*! That is so cool!"

"*And?*" Jamie said.

Kelsie continued, "And he hasn't said much about it. The last entry had a few references to it. There were a lot of emotions and all that, but not much information. But I'm not totally done dyna-reading."

"Is that why you threw the calipers?" I asked.

"No. I was frustrated because there's another riddle just to turn it on."

There were several loud moans, the most obnoxious from John.

"Well, what is it?" I asked excitedly.

Kelsie took a deep breath. "We have to use the *elements of caution* to turn it on."

We all turned to John. "What? Don't look at me," he said putting up his hands.

"You're the one who has been working on the teleportal," I said.

"I know, but I've just been rewiring it, not figuring stuff out."

"Any idea what the *elements of caution* could mean?" I asked.

"Whatever, man. It probably means: *the cake is a lie,* or something" he retorted.

"John, I'm serious."

"I don't know. There's just that blank keyboard—and that sign."

"Well," Kelsie huffed, "finish rewiring it and then Matt and I at least will figure this out."

12

Turning on the Teleportal

After a few more days, John finished rewiring the teleportal. John, Jamie, Thomas, Ari and I were waiting around during study hall. Kelsie finally came in, "Why are all the younger kids wearing electrical cable belts?"

John lifted up his shirt, revealing three or four strands of cable wrapped around his waist—and some of his hairy stomach.

Kelsie averted her eyes, "Eyew! John, don't ever do that again. Ever!" She came over to join us. "So, you've started a fashion trend?"

"Dude, I don't know. It's not my fault they worship me," he said smugly.

"How do you …" she apparently didn't know what to say and instead brought her fingertips together three times.

"Keep the belt tight?" John offered. "I zip tie them. Do you want to see?"

"No!" she screeched.

"Come on, you guys! Let's do this," I snapped.

Like the team that worked on the first Atomic bomb during World War II, we had no idea what was actually going to happen. As scientists have found out the hard way time and time again, we simply had to switch everything on and hope for the best.

Maybe things like that were coursing through everyone else's minds, because they all looked as nervous as I felt. John was visibly shaking. His hands were quivering as he bit his nails and ran his fingers through his hair. Ari kept clenching his jaw, his cheekbones pulsating. He kept wiping his hands on the front of his jeans. Jamie and Thomas were pacing back and forth so much that I finally yelled at them to cut

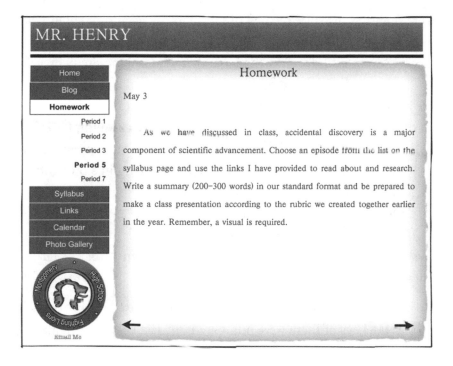

MR. HENRY

Home	
Blog	
Homework	
Period 1	
Period 2	
Period 3	
Period 5	
Period 7	
Syllabus	
Links	
Calendar	
Photo Gallery	

Homework

May 3

As we have discussed in class, accidental discovery is a major component of scientific advancement. Choose an episode from the list on the syllabus page and use the links I have provided to read about and research. Write a summary (200–300 words) in our standard format and be prepared to make a class presentation according to the rubric we created together earlier in the year. Remember, a visual is required.

Email Me

it out. Kelsie had her eyes closed and was slowing down her breathing using some Tai Chi exercise or something.

We weren't exactly face to face with death, but I have to say that this was the closest I had ever been. I think that this was true for everyone except Ari. When he was eleven, he was in a car accident and almost died along with his parents. He described the whole incident: how the car hit the railing, went up an embankment and slammed into a tree—all in slow motion. He remembers thinking: "Hmm, that's weird. That's my dad flying out the windshield. That's going to be really bad. Ohhh, that's the seat in front of me coming toward my face. This will hurt …"

I don't know why time seems to slow down in those situations. Maybe it's a survival instinct or something. But I suddenly understood what Ari said a lot better. John looked around and then nodded, but it looked like he was in a pool of molasses or something. His motions were slow and his voice seemed distant. I'm not even sure about what he said that day, but the rest of us must have agreed because then John slowly—ever so slowly—moved his hands and pulled the circuit breaker.

I was suddenly aware of a loud humming and a strange buzz in the air. It felt like there were millions of mosquitoes swarming around all of us, but the noise was more consistent, more constant. I was lost in the sound, trying to comprehend what exactly was happening. My head was swimming and everyone else was moving in slow motion. And then I blinked and life suddenly seemed to catch up.

"Okay," John concluded, "my part is done."

Once the power was delivered to the teleportal, the next problem was that we had to somehow switch on the machine itself. As far as I knew, we had three clues:

1. The keyboard attached to the teleportal with 120 keys (only 118 keys of them were numbered; the last two were blank).

2. The sign above the teleportal read CAUTION.

3. The phrase from the journal: "Use the elements of caution."

The keys were arranged in three tiers with forty keys each. They were numbered 1 to 118 with two blank keys at the very end. What was significant about the number 118?

Like vocabulary words, some numbers are memorable and some are not, and some seem inherently cooler than others. Like most things, numbers are interesting (or not) based on their traits or properties.

For example, if you think a movie is interesting, not only do you like it, you like it for various reasons, even if you don't consciously know what those reasons are. You might like the acting, the plot, the special effects, the action, the camera work, or any of numerous other things about it. If you like a particular painting, then there must be some trait in the colors, the textures, the patterns or something that you enjoy viewing.

Numbers are the same way. They have traits about them that make them different from other numbers, traits that even make them unique, just like people. Mathematicians play around with numbers, just as some people like to doodle or draw. That's how many of these properties or traits are discovered. It may seem weird to play with numbers, but it's really not any different than randomly doodling on a piece of paper to create weird swirls and pictures.

An example of a simple interesting number is 2 because it has a lot of remarkable traits. It is the first even counting number, and is the only even number that is prime. Two things make a pair, and pairs are how most living things reproduce. Two shows up in basic computing, namely binary code.

Eleven has some interesting traits as well. It is the first number that cannot be counted on a human hand. It is the first counting number that is made from the same digit twice (other numbers like that include 22, 33, 44, 55 and so on). Eleven is the supposed number of space-time dimensions in string theory. There are eleven players on the field at a time in soccer, field hockey, football and cricket. The "eleventh hour" refers to the time when it is almost too late to get something done.

Another interesting number is 153. One of its traits is that it is the sum of its digits cubed: $1^3 + 5^3 + 3^3 = 1 + 125 + 27 = 153$. It is also the sum of the first five factorials: $1! + 2! + 3! + 4! + 5! = 1 + 2 + 6 + 24 + 120 = 153$. It is also the sum of the first seventeen positive integers. Curiously, in the Bible it is the number of fish that Jesus and the Apostles caught. By the way, 153 is the Dewey Decimal System number for "mental processes and intelligence."

There are imaginary numbers, which are really cool. The most famous is probably the square root of negative one. It's imaginary because squares can't be negative. The reason is a square root is asking for a number times itself. Whether the number is positive or negative, when it is multiplied, the product will be always be positive. You might wonder what good an imaginary number might be, but there are applications. Imaginary numbers are used in fields like electrical engineering, control theory, and quantum mechanics, where they help to explain certain phenomena in things like electrical circuits.

This reminds me of something called the Interesting Number Paradox. This mess-with-your-mind paradox states: *the first number that is uninteresting becomes interesting for that reason.* That probably needs repeating: *the first number that is uninteresting becomes interesting for that reason.*

So, take 629 as the example. Let's call this number uninteresting because there's nothing apparently special about it. But the very fact that it is uninteresting *now* makes it interesting. In other words, it is interesting because it's not otherwise interesting. The paradox gets worse, though. Let's say that all numbers are either interesting or dull. Start counting and determine whether the number is interesting or not. One is interesting because it's the first positive integer. Two is interesting, for example, because it's the only even prime number. Three is interesting because, among other things, it's the first odd prime number. Four is interesting because it's the first composite number, meaning it has factors besides one and itself. Five is interesting because it's how many fingers and toes we have on a hand or foot.

You get the idea. At some point you'll encounter a number that seems to be dull, or uninteresting because there's nothing noticeable or exceptional about it. But then it becomes interesting, because it was the first dull number that you encountered. So move on to the next number. It will also seem to be dull, but will become interesting because it is *now* the first number that is apparently dull. This could make *every* number interesting!

Anyway, I kept thinking about the phrase from the journal and how it might tie everything together. What would the elements of caution be? Why was the word "elements" plural? And what was with the number 118?

Here are a few things I collected about 118:

- There were 118 episodes of Kojak, one of mom's favorite shows.
- 118 A.D. is supposedly when the Roman Forum was completed.
- 118 was the highest grade Kelsie ever got on a test or quiz.
- Ununoctium has an atomic number of 118.
- In binary code: 1110110.
- Prime factors: 2 x 59.
- Roman numerals: CXVIII.
- A dime has 118 grooves on its edge.
- In the Dewey Decimal System 118 is the number for "force and energy" ~very interesting!

Then I started thinking: *What are the elements of caution?* In other words, what is caution made of? One element must be expectation. Without expectations you wouldn't be cautious. Although this was pretty cool to think about, I had no idea how to translate it into numbers. I thought of pressing the numbers for spelling out words like expectation: 5 for 'e', 24 for 'x' and so on. I considered spelling out "element" and then I tried spelling out "caution." But these things wouldn't explain why there were 118 buttons. Just spelling out words this way would only require 26 buttons, one for each letter of the alphabet. So what would need 118 buttons?

"So, what do you think, John?" I asked.

"Um … maybe it really said *elephants of caution* and we have to bring elephants in here."

"I'm serious, John."

"Okay. Um, elements of caution … Do you think it means earth, air, fire and water elements? Or the periodic table …"

"We said that already."

"… *of* caution."

"The periodic table *of* caution? What would that even mean?" I asked.

"Well, what about the elements themselves? Many of them are dangerous," John suggested.

I pulled out my phone and opened the periodic table app. "Well, mercury is poisonous and …"

"Uh, hello," John quipped, "what about all of the radioactive ones?"

"I was getting to those, John."

"Dude, you start with *Mercury*?"

"John, that's not helping."

"Okay, sorry."

"Maybe we have to press the buttons for the dangerous elements," I suggested.

"Actually," John said, "if you think, about it all of the elements need caution."

"What do you mean?"

"Well, the obvious ones are radioactive or poisonous, but the others need some caution as well. Too much calcium in your body and you could die or …"

"Die?"

"Well, calcification in the soft tissues could cause complications. It has something to do with balancing calcium and phosphorus." He looked more serious than I remember him being in a long time. "What?" he said finally. "I pay attention in honors biology."

"It's not that. I was just thinking: What numbers should we press then?"

"Dude, I don't know. Everything on the periodic table can potentially be dangerous. Hydrogen is obvious. Zinc, copper, iron, potassium, magnesium are all potentially dangerous. Take sodium, for example. Too much sodium attracts and retains water, which increases your blood volume and stresses your heart. If pretty much anything is out of balance, we can die."

"Exaggerate much?"

"I know, but I'm making a point that they all need caution."

"What about something like helium?"

John started laughing, "You could suck too much into your lungs, make high pitched squeaky noises like The Chipmunks and die laughing!"

After a weird picture of imagining the six of us dying horrible laughing deaths as Alvin, Simon, Theodore and a few new chipmunks flashed through my mind, I realized that John was right. The phrase *of caution*, wasn't much of a clue, but thinking about it as a physical warning of some kind wasn't helping either.

The next thing I knew John was talking out loud, "Elements *with* caution, elements *of* caution …"

"That's it!" I yelled. "The Abbreviations! Ca for calcium. U for uranium, Ti for titanium, O for oxygen and N for nitrogen." We pressed the corresponding numbers: 20 for calcium, 92 for uranium, 22 for titanium, 8 for oxygen and 7 for nitrogen. I was so excited about our discovery that when the machine didn't turn on, I was completely shocked.

These were the elements of my shock: Sulfur, Hydrogen, Oxygen, Carbon, and K for Potassium. And I was disappointed. And a bit depressed. And hungry, by the way. I wanted some pizza. P-I-Z-Z-A, hold the Phosphorous, Iodine and whatever undiscovered elements "Z" and "A" could abbreviate.

We were missing something, but what?

PART TWO

A MatheMATTical Adventure
A MatheMATTical Adventure
A MatheMATTical Adventure
A MatheMATTical Adventure
A MatheMATTical Adventure
A MatheMATTical Adventure

13

What a Trip

We spent the next few days at lunch discussing the teleportal and what might be on the other side. One of the first ideas was that the portal was for time traveling or teleporting. As is the case with people, the more days that passed, the more our stories became exaggerated. John figured that this was a Stargate from the movie and television series. Ari dreamed that it would take us into the Star Wars universe, while Jamie began insisting that the Star Trek universe would be better. Kelsie just rolled her eyes. I remained scientifically reserved and didn't form an opinion.

Thomas and Jamie were adamant that we shouldn't even try the teleportal. They were worried about all of the horror stories from science fiction stories and role-playing games: showing up in the middle of the ocean, teleporting into a wall, being biologically scrambled and arriving with three arms, genetic mix-ups and much more.

John was actually more worried about what might come through the other end into our world. Maybe the Maglio person martyred himself to save the world from human-enslaving aliens. Ari was doubtful the thing was even going to work. Kelsie and I seemed like we were the only ones who were excited about the teleportal. She was up for an adventure—and keep in mind, her idea of adventure is wilderness survival school and eating lichen "potato" chips.

At the end of lunch one day I found myself alone with Kelsie, the others way ahead of us in the hall. I realized then that she had slowed our pace down on purpose. She had her jacket on and her backpack slung over her shoulders. She handed me another pack when I got closer. "Here, Matt, take this would you?" I grabbed it by the top handle and it was so heavy I had to lower it to the ground.

"What's going on?" I asked.

"I think we should go get Maglio. Come on, Matt." She motioned toward the boys' bathroom.

"You mean … right now?"

"Yeah. I'm all geared up. I say we go now, just the two of us," she said pulling my arm.

"But I'm not even prepared. I only have my iPad," I said holding it out in front of me.

"Don't worry, I have enough gear for both of us," she pointed at the bag I held and then hitched her thumb toward her backpack. "You can leave your backpack in the lab."

"Just us?"

"We're the only two who aren't bickering and driving each other crazy."

"But what if …"

"No *what ifs*, Matt."

"But it could be dangerous. We should wait for the others before …"

"Don't worry," she winked at me. "I can take care of the two of us. And anyway, Ari already left on the basketball bus, Thomas and Jamie are just hopeless and we need John to stay on this side in case something happens. Besides, if something *really* bad happened, I might not be able to take care of everyone."

I was about to protest further, but there was something irresistible about the way she was looking at me, and the wink was too much. I suddenly didn't care that we might be going off on a dangerous mission. At least we'd be together. Alone.

"Come on, Matt." She put her arm through mine and steered me toward the bathroom door.

Let me pause here to say that I have on several occasions thought about asking Kelsie out on a date. Officially, that is. We have been best friends ever since I can remember, so it has been a little awkward. We have been to the movies and stuff plenty of times, just the two of us, but they were never dates exactly. In the past few months I couldn't even tell you what some of the movies were about that we had gone to. I was too busy being confused sitting next to her. I thought about putting my arm around her, trying to hold her hand, putting my hand on her leg. I thought about trying to move closer and looking at her in a "more than friends" kind of way. But I never followed through with anything more than squirming uncomfortably and later wondering what the movies were about that we saw.

So when she put her arm through mine, all of those times we were together flashed through my mind. For several dozen steps, I felt like she was my girlfriend and that we were racing off to the great unknown, which we were. The next thing I knew we were standing in front of the teleportal.

"We still have to figure out how to turn it on. I know we're close." She circled around the machine. "What are we missing? I thought we had it before, when we spelled out caution."

"That's it, Kelsie! You can spell caution differently!"

"Wow, Matt, you really are a genius, aren't you?"

"No, I mean it!"

"Are you trying to tell me there's a British way to spell it or something?"

"No … wait; what? No, with the periodic table. We used Ca for calcium and U for uranium. But we could have tried C for carbon and Au for gold!" She slapped me high five as we leaned over the keyboard. This time we pressed the corresponding numbers: 6, 79, 22, 8, 7. There was a loud science fictiony sound, as Kelsie put it, followed by a buzzing and then the floor started vibrating. The machine was on and ready to go.

Kelsie's touch was lingering on my arm when I realized she was no longer next to me and she was calling me. "Matt! Hey, Matt plug that in, I'm going to throw the breaker." She sounded so confident, so sure of herself that I just went through the motions and the next thing I knew she was talking to me, "Are you ready?"

"Wait! Something's been bothering me about all of this."

"You want to talk about it *now*?" she said, glowering at me.

"Well … yeah. I feel like the codes have been a little too easy …"

"Too easy? Some of them took *you* days."

"I know, but any of them could have been encrypted with strings of random numbers and letters."

"So …"

"Then only people with those strings or passkeys would ever be able to get in. That's how we safely buy stuff online."

"Then obviously Maglio wanted people to break these codes. That's what I have been trying to tell you. I think Maglio knew that he was doing something dangerous and left a trail of breadcrumbs behind. He *wanted* people to find him."

"But why?"

"I don't know. I read and dyna-read most of his journal entries, and Maglio was angry, and hurt and isolated. No one respected him or his work. I think he tested the teleportal on himself. Now, he's stuck or in trouble."

"He could be dead. He could have teleported into the middle of a black hole or something. This could kill *us*, Kelsie. You do realize that, right?"

"I know, but according to his journal entries, he brought back samples of things that he found on the other side. In one he described a new species of tree. He made the trip several times, and then I think the circuits got fried and he couldn't return."

"But John fixed all that. Maglio might be able to return on his own."

"But if he's been there a long time, he's probably given up. We need to go there and bring him back."

I wanted to argue more. I really did. I wanted to explain to Kelsie that trained people like Marines should be the ones for a search and rescue operation. I wanted to point out that we might go through and fry the circuits and then John would have to fix it again. Then maybe

Thomas and Jamie would come looking for us, fry the circuit, and so on until we all disappeared in separate small groups like in horror movies. I wanted to point out that we were under-prepared, under-trained, and under-just about everything.

But she looked at me in a certain way, and my resolve crumbled.

Plus I thought, *If I am going to die, at least I'm with Kelsie.*

She pulled the circuit breaker, and then put her arm magically through mine. I could smell her apple-scented shampoo as she stepped closer. "Ready?" she said. I nodded.

"Here we go," she said, tugging my arm.

"Wait!" I said in a sudden panic. "Do you think this will fry my iPad?"

She looked at the teleportal and finally nodded, "Yeah, you should probably leave it here." When I let out a deep breath, she said, "Don't worry, I have my old iPhone."

"All right," I said, setting my iPad down.

"Ready?" she said. I nodded.

Leeroy Jenkins! crossed through my mind as we stepped into the teleportal.

I will try to explain what happened to us when we stepped through that portal. Imagine somehow being pulled through a garden hose. I mean, imagine the whole thing, like picturing a cartoon character. First imagine being squeezed and crushed one piece at a time in order to even fit through something as narrow as a garden hose. Imagine the intense pain just in your arm alone, say, as it's being run over by a huge truck. Then multiply that feeling times every part of your body. Add to that tumbling haphazardly around the inside of the garden hose, bouncing off the walls and landing awkwardly on your ear, and that's about what I felt like for an instant as we "traveled" somewhere. Okay, I exaggerate a little, but it hurt.

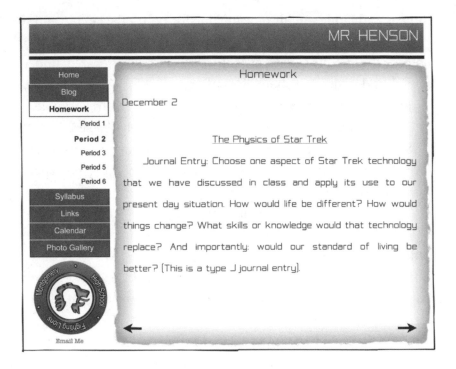

Imagine a single point in space—you'll have to imagine it, since just a single point is hypothetical. If you were to try to draw one, for example, and made a tiny dot on a piece of paper, then this is FAR from a single point in space. Look at what you did under a microscope and you'll see what I mean. Your little point is made up of smaller points, which in turn are made up of smaller points and so on. Another way of thinking about this is that each dot that you create has a radius and that radius has a midpoint, which has a radius and midpoint, which has a radius and midpoint. This is why a single point can be thought of as having zero dimensions.

Now take that point and move it to a new location. Now you have a line segment, which now has one dimension. What's interesting is that this might even be how the number one was formed. Think of it as a line segment. And that line segment symbolizes the first dimension.

Now take that line segment and move it. You have just created a square or rectangle, which has an area. This is the second dimension, and we use square units to measure this area. Notice the "2" in the units squared, such as 32 m², or thirty-two meters squared.

Now take the square and move it through space. You now have a cube in three dimensions, which is the first time we can call it an actual object. We use cubic units to measure this volume. Notice the "3" in the units cubed, such as 81 m³, or eighty-one cubic meters.

Move the entire cube through space and you have the fourth dimension: time—or more accurately, time/space. We don't use spatial units to measure this, so there's no little "4" like the others. But we do introduce something new in the units: time. Not only do we have seconds, minutes, hours, years, and centuries, but we also have distance & time units. For example, we have measurements like miles *per* hour or kilometers *per* second.

Now things get weird, which is why I bring it up here. From what I have read, the fifth dimension has not been fully defined or agreed upon by experts. Generally, though, the fifth dimension is *beyond* time. This has various meanings in science fiction, such as bending time, going backwards in time, or skipping around in time. But these things bring up weird issues, such as arriving somewhere before you even start, or traveling faster than the speed of light, which has its own issues.

You can actually continue the drawings to help understand what might happen. The spatial dimensions go from the dot to the cube, and then the fourth dimension is that cube moving through time. For our normal living, time is an arrow, or a ray, that moves only in one direction. We can't go backwards in time, for example. But imagine if there's more than one ray. Imagine if every time you have a decision to make that there are several rays. For example, you have to decide what color car you want. Path "A" is blue, path "B" is red and "C" is lime green (I love the thought of a lime green car). After choosing one of the cars, then your arrow of time continues and you can't go back.

But imagine that the other two arrows still exist. If you could travel *sideways* to one of the other arrows, then you'd suddenly have a different car. The space in between the arrows could be regarded as different dimensions. There are numerous problems with such thoughts, though. If you could ever travel sideways and/or backwards, then you confront one of the classic paradoxes. What if you went "back" to a time/place where your grandfather or grandmother died before giving birth to one of your parents? Would you cease to exist? Is that path simply not possible for you to travel to? What if you went sideways to a time/place where you had already died? Or what would happen if you met up with yourself? The other big issue is that there would be an infinite number of these rays and infinity is troublesome too. But that's for another time.

I have to say "traveled" through the portal because I don't really know what happened to us. I don't know if we "fell" through space, or shifted dimensions, whether space "folded" and we were suddenly elsewhere, or whether a place came to us. All I can say is that I was highly disoriented, dizzy and confused. I'll admit that I practically felt like crying. Hey, I'm man enough to admit it.

Kelsie is the one who brought me around by not so gently smacking me on the top of my head. "Matt, get up you baby. That didn't hurt that much."

I've heard that women have a higher threshold for pain than men do. Now I had my own physical proof. I could barely move. My joints hurt. My bones hurt. My skin felt like it was burning. A headache throbbed behind my eyes. My stomach was knotted and clenched as if I had a horrible stomach ache. My fingers cracked when I moved them. And, well, I could go on listing everything that pained me, but the point is that while Kelsie was *standing* above me looking around, I just wanted to curl up and die.

"Matt, you've got to see this!"

All I could see clearly were her black hiking boots and her light blue socks with white fluffy trim. They were pretty cool, actually. Why don't guys wear frilly things?

Kelsie reached down, grabbed my hand, and pulled me up. My arm deadened. Suddenly, she was trying to pull all of my weight. I felt just this side of useless. She caught her balance, grabbed me with her other hand and helped me to my feet. She's surprisingly strong.

14

Senses

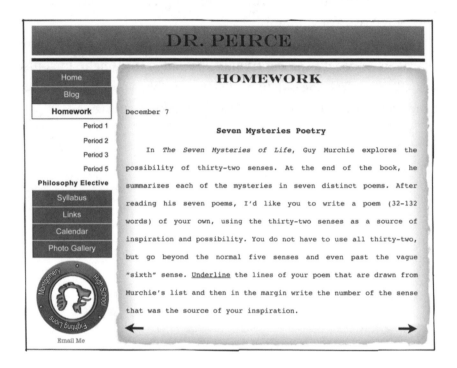

Home
Blog
Homework
Period 1
Period 2
Period 3
Period 5
Philosophy Elective
Syllabus
Links
Calendar
Photo Gallery

Email Me

DR. PEIRCE

HOMEWORK

December 7

Seven Mysteries Poetry

In *The Seven Mysteries of Life*, Guy Murchie explores the possibility of thirty-two senses. At the end of the book, he summarizes each of the mysteries in seven distinct poems. After reading his seven poems, I'd like you to write a poem (32-132 words) of your own, using the thirty-two senses as a source of inspiration and possibility. You do not have to use all thirty-two, but go beyond the normal five senses and even past the vague "sixth" sense. Underline the lines of your poem that are drawn from Murchie's list and then in the margin write the number of the sense that was the source of your inspiration.

After standing up, I noticed several things as my perspective became clearer. One was that the air was fresh and a cool breeze was soothing my skin. I could smell Kelsie's apple-scented shampoo again. Another was that gravity felt about the same. People talk about a sixth sense all the time, but there are a number of sources out there that talk about more. I've never been satisfied that there are only six, mainly because six has never seemed like a completion to me. Obviously, the numbers seven, eight, and nine follow. Think about seven days in a week, eight notes in an octave, or nine months in a human pregnancy.

Rudolf Steiner suggested that there are 12 senses, including *balance, life,* and *movement.* In an amazing book called T*he Seven Mysteries of Life,* Guy Murchie actually puts forth the idea that there are *32* senses, including *radiation.* These things have been on my mind for a while now, especially because they involve numbers. Why 12, for example? Are those related somehow to astrology, or to our concept of a dozen, or maybe the months of the year? Why 32? Is there some thing special about that number? Why not 33 senses, or 34? Why not 31 or 29? And perhaps most of all: why aren't more people talking about more senses? I can't be the only one interested.

Anyway, maybe because those things were on my mind, or maybe my senses were heightened from being in such a weird circumstance, I quickly had several very strong feelings or senses—I'm not sure what to call them, actually. One was that this place was *not full of life.* I realized that, despite being in an open field full of grass, trees and occasional bushes, there didn't seem to be much else. I wasn't sure if this sense was like human radar and I wasn't detecting other people, or what, but the place seemed rather empty.

The next was a sense of *imbalance.* This one is hard to describe, but imagine walking on a sidewalk at an angle close to 45 degrees. That's what it felt like. One side of my body was shortened and awkward and the other was constantly searching for solid ground. My head felt lopsided, the left side significantly heavier, like it was made of iron or something. The right side was light and felt like it was trying to float away. For those who have been on a boat before, it felt like I was on a poorly designed deck that was not parallel to the water.

The next sense was *wrongness.* If you have ever walked into a room where people have just been fighting, or accidentally walked in on two people breaking up or something, then you can understand this one. I felt awkward and out of place, maybe even a little embarrassed. Not quite the nightmare level of finding yourself at school wearing just your underwear, but certainly close. I actually checked to make sure I had all my clothes on. I did.

I'm not sure what to call the last sense, but maybe this is what Guy Murchie described as the sense of *radiation.* There's really no other way I can explain it. It was not the wind tickling my skin. It was not the sense of excitement (hmmm, is that really a sense?). It was not the fact that I was alone with Kelsie in a weird, possibly alien landscape. No, this was something completely different. If you've ever put your hand next to the television screen and felt the static electricity, this was close. But it was not just on my hand; it was everywhere. Imagine that static feeling all over your hands, arms, face, your *entire* body. It made me itch all over.

These senses flooded over me, and in my already overwhelmed condition, I nearly fell over. Once again, it was Kelsie who stayed strong and helped hold me up. "Do you *feel* that?" I asked her.

"Feel what exactly? I'm feeling a lot of things right now," she said squinting into the distance.

"All of it. Everything. The weirdness, the tingling, the …"

"Emptiness." She let go of my arm, checked to see that I wouldn't fall over (I didn't), and then she spun slowly around in a circle. "That's what I feel. Empty—or almost empty."

She took out her iPhone and looked at the screen. "Oh well," she said, "it was worth a shot." I reached for mine too, and then realized I had left it in my jacket pocket back in my locker.

"No signal?"

"Nope." She started to put it away and then stopped. She brought the phone up closer to her face. "That's weird."

"What?"

"I have your number on speed dial on my screen and the eights are flickering."

"What do you mean?"

"Look!" she said shoving the phone at me.

I saw my phone avatar on her screen and, sure enough, the three eights were flickering 1888, which I have always liked, since it has the cool property of being the longest year to date written as Roman numerals (MMCCCLXXXVIII), something that won't be broken until the year 2388. "Hold on," I said taking her phone. I scrolled through her address book. "They're all flickering. All the eights are glitching or something."

"What's it mean?"

"Um, I'm not sure. Maybe traveling through the portal …" I shook the phone as hard as I could.

"Matt!"

"Just trying to fix it." I looked at the screen and my phone number was back to normal. "See, it worked!" I laughed.

Kelsie grabbed her phone back, took a long look at it and then put it back in her pocket. "Come on, Matt, let's go over to that pin oak." She pointed to the only tree within one hundred and ten yards or so of us. That's when I looked around at all of the trees and noticed they were in various seasonal stages. Some of the them were budding like in spring, some had the green leaves of summer, while on others the leaves were the reds, oranges and yellows of fall. Still other trees had no leaves, probably a sign of winter, although I don't know enough about trees to tell if they were truly dead or not.

"Kelsie, look at all of the trees."

"I know," is all she said.

"Any ideas where we are?"

"You mean besides being on the other side of the teleportal? How am I supposed to know?" She circled very slowly looking for something. "Come on," she said.

"Why there?"

"Well, there's no portal back and we have to do something." I looked around and Kelsie was right. No portal. Her words echoed in my head. How were we going to get back? "No portal?" I mumbled.

"What did you expect, Matt? We're on an adventure!"

"I know, but how are we …"

"Look, we have to find Maglio and then he can help us get back—just like in all the movies."

She was right, of course. In the movies, portals always close behind, trapping the main characters somewhere. There wouldn't be much excitement if Maglio had simply been waiting here, and we just grabbed him and jumped back through.

Kelsie steered us toward the large oak tree. She jogged a little ahead and then slowed down close to the tree. Her head was focused on the ground and she started walking in circles again studying the ground.

"What are you looking for?"

"Tracks. Signs." She had gone to wilderness camp one summer and had learned about tracking. I'd like to think that was why she was so calm and collected under these circumstances and that I wasn't. She simply had training and I didn't. It had nothing at all to do with being a wimp or anything.

I looked at the ground and didn't see anything special—just grass, leaves and twigs. Suddenly she tilted her head and started walking slowly down a little incline leading away from the tree. "There are tracks here. Look," she said pointing to a patch of grass and leaves.

I looked but I have to say I didn't see anything until she pointed to a specific spot—and even then she had to outline it with her finger. "Is that a footprint?" I asked.

She nodded. "It could be Maglio's. I don't know for sure. Hold on. Step on the grass right next to it, full weight. Okay, now back up a second." She knelt down even closer to the ground. "Okay, I want to see how the grass reacts."

"How the grass reacts?"

"Watch."

I watched.

"See how it's springing back a bit?"

"Um, I think so." I'm not sure that I did.

"We can compare your track to these and maybe get an idea of how old they are. What size shoe are you, Matt? Nine?"

"Nine and half, actually."

"Okay, so these are size ten. Probably a man, although not necessarily."

"Is there a way to tell the gender of a track?"

"From what I've heard, yes. But I don't have anywhere near enough experience."

"How is it supposed to work?"

"I'm not sure, but basically the idea behind tracking is that everything leaves a unique trail or mark, just like fingerprints. So, men and women leave different types of tracks."

"But you can't tell gender from fingerprints."

"I know. But I told you: I don't have enough experience." She suddenly sprang up to a standing position. "Anyway, I can't wait to take more classes!" She circled around the track a few times and then changed her gaze farther down the slope. "Come on!"

We walked for a while and stopped periodically so Kelsie could examine the ground. She assured me that she was following a set of tracks, but I could only see what she was talking about once in a while. After a lot of conversation where we seemed to purposely avoid talking about school and family, she suddenly swung her backpack off her shoulders and sat down. "You hungry?"

"A little." I slid the backpack she had given me off my shoulders. "What's in this thing? It's heavy."

"Gear. Mostly water bottles and power bars."

I nodded absent-mindedly.

"I have some beef jerky," she said as she opened one of the side pockets. "Do you want some?"

"Yes!" I practically yelled. She handed me a piece.

"Geez, Matt! Hungry?"

"I didn't have breakfast and now must be … well, it feels way past lunch. What time is it, anyway?" I chomped into the jerky and savored the salty goodness.

"Well, based on the sun, I would say maybe four o'clock—although I can't be sure." She took a bite of her jerky. "We have a few more hours until it starts getting dark."

"Then what?"

"I'll make a fire and we'll set up camp," she stated, taking another bite.

We sat there for a while and talked about what the other members of the Company were probably doing. I stirred my fingers through the grass until I noticed a patch of clover. I bent closer, "Hey, cool," I said, pulling one up, "a four-leaf clover." Kelsie looked up and smiled. I felt my cheeks blush a little. I hope she didn't notice.

"Make a wish," she said laughing, "it must be your lucky day."

"I wish … to be on an adventure with Kelsie …"

"Hey, wait, Matt! I found one, too," she said, holding up another four-leaf clover. Then she paused, her head tilting to one side. "Uh, Matt, they're all four-leaf clovers."

I speared my fingers through the clover, isolating several at a time between my fingers. She was right; they were all four-leaved. "What do you think it means?"

"I'm not sure," she said.

"Maybe they're genetically modified."

"From what?"

"I don't know. That weird tingling could be radiation or something."

"You mean nuc-u-ler radiation?" she said, mispronouncing nuclear in a George W. Bush imitation.

"Yeah," I laughed, "something like that."

"Hold on." She got up and scanned the area around the tree. Then she wandered a bit, her eyes fixed on the ground in a wide view. "They're *all* four-leaf clovers. Every single one of them!"

"Really?" I believed her, but crawled around on the ground to verify anyway. It seemed too weird to be possible. She was right.

"Hey, maybe here we have to find a three-leaf clover!" She got excited then and started running between patches of clover. I crawled around faster. Suddenly we were in a race to get a lucky magical three-leaf clover of ... um, luck.

"Did you find one?" she said finally.

"Nope. Just fours. We should bring these back and sell them or something!"

Kelsie ignored me. "So it's threes here?" she asked, pushing the stick into the ground a bit more.

"I guess so."

"C'mon *MatheMatt*! That's all you have to say is *I guess so*?"

"Can I say *I guess so* again?"

"No! Seriously Matt, you must be thinking about something " I actually was thinking about being stuck here forever with her and how I didn't think it would be that terrible. I didn't want to say anything like that though, at least not right now.

I was about to say something about the clover when I spotted an ant milling around on the ground. I followed it for a bit, and when it crossed paths with another ant, I followed that one for a while too. They seem like random patterns, but, of course, they're not random at all. They just *seem random* to people. From the little I know about ants, they're actually moving very purposefully following chemical trails. I was engrossed watching them when I noticed something. "Uh, Kelsie, check out these ants."

"What about them?"

"How many legs do ants usually have?"

"Six, like all insects. They're arthropods and ... wait, why are you asking?" She suddenly looked a bit concerned.

"Well, these have seven."

"Seven? Are you sure those aren't antennae because ..."

"Yeah, I'm sure," I said, picking one up on a stick for a closer look. "See," I said holding it out to her. Sure enough, it had seven legs. It had its six "normal" legs, two antennae and a seventh leg coming out from the back of its abdomen.

"It's weird that it's not symmetrical. Maybe it had eight legs and lost one," she said.

"Wouldn't that make it a spider?"

"Well, technically ... but this is almost certainly an ant."

"An ant with *seven* legs?"

"Yeah." She stood up suddenly and walked closer to the tree. Then she got on her hands and knees very close to the ground. "They all have seven legs. Every one of them."

"This is *so* weird," I said, not really knowing what else to say.

Kelsie crawled back over to the tracks. "Come on," she said standing up, "Let's keep going. Hopefully, Maglio has some answers."

We continued finding unusual examples of things, mainly because Kelsie had so much knowledge about the outdoors. We came across plants that she could identify and in each case, there was something numerically unexpected about them. We found a few white pine trees that had needles in clusters of eight instead of five; some wild strawberries (which were very bitter and not at all like the ones in a store) with a nine-part flower instead of five; and apples (which were pretty tasty) that Kelsie cut open that had four-part symmetry instead of five.

Normally, I would just think that we had been finding different species, or that the seven-legged ants were just a strange turn of evolution or something.

But the snowflakes definitely convinced me that something was wrong.

We were still following the tracks when the temperature dropped drastically like we had walked into a giant freezer or something.

"Whoa!" I said, shivering, "Kelsie, are you giving me the cold shoulder?"

"What?" she responded, slinging her backpack down.

"Never mind."

Kelsie was prepared, of course. As it turns out, she saved our lives, although I get some bragging rights too. But that part of the story is coming.

I was amazed at everything she had packed in there. The first to come out was a coffee can, followed by a small jar of peanut butter, then a blue poncho or something, and a handful of wooden sticks. After that, she pulled out a few very small, fuzzy bundles that were rolled up tight and wrapped several times by a thin bright green cord.

"Is that dental floss?" I asked, laughing. The thought of flossing my teeth seemed ridiculous right now.

"Yep," she said, unwrapping a blue and purple bundle. I watched as they unfolded into long-sleeved fleece shirts. It must have taken her hours just to roll up everything in her bag.

"Let me guess: survival school?"

"Yeah, floss is good for all kinds of things. You can use it for fishing line. You can tie things together and make snares with it. Not as universal as duct tape, but pretty close." She shook the blue shirt so that it unfurled completely and held it up in front of me. "This should actually fit you."

I pulled it on and I'm a little embarrassed to say that her shirt did fit me. I have to start working out when we get back … or if we get back.

She pulled on the purple shirt and then put the other stuff away. She saved the coffee can for last. "What's in the can?"

She opened the lid and stirred the contents around with her hand so I could see most of it. "Survival stuff. Some duct tape, sunblock, needles, flint and steel, matches, Ziploc bags, some garbage bags, stuff like that."

"Garbage bags? I thought you recycled everything." She didn't laugh at my lame joke, maybe because she was in survival mode or something.

She hardly missed a beat. "They're good for quick shelters, emergency clothing and even carrying water and stuff." She continued showing me stuff. "Some power bars, my pocket knife, a compass … hmmm, I actually want that … a candle, a whistle, some water purification tablets and a few other things."

"What's the mirror for?" I asked as she looped the string to the compass around her neck.

"Signaling mainly, although if you grow a beard while we're here, you can use it to shave." Ouch. She was making a joke about my baby face. It would take me decades to grow a beard.

"Come on," she said, "let's keep moving."

With the fleece on, I was much warmer and shivered less. The first snowflakes that fell were the huge fluffy clusters that swirl around even in a gentle breeze. I was immediately caught by the vortexes and swirls in the wind. Some of the flakes got caught on the fleece and I couldn't help but stare at them.

15

Fifty-Seven

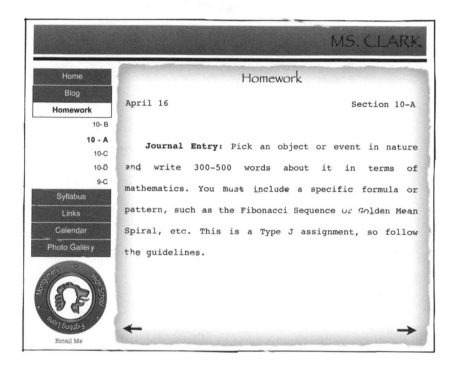

MS. CLARK

Home

Blog

Homework

10- B

10 - A

10-C

10-D

9-C

Syllabus

Links

Calendar

Photo Gallery

Email Me

Homework

April 16   Section 10-A

   Journal Entry: Pick an object or event in nature and write 300-500 words about it in terms of mathematics. You must include a specific formula or pattern, such as the Fibonacci Sequence or Golden Mean Spiral, etc. This is a Type J assignment, so follow the guidelines.

As someone who loves math and structure and order, snowflakes—or more technically, snow crystals—have always fascinated me. Although there are more than thirty different types of snowflakes, almost all of them are crystals in shapes based on six-sided forms. There are six-sided columns, hexagonal plates, six-sided bullet shapes, fern-like crystals with six arms, and even star-shaped flakes. Sometimes, there are twelve-sided flakes, but I think they are technically just two sixes merged together. There is a very rare type that seems triangular, but even those pretty much have six sides and don't look all that different from ordinary crystals. It's actually amazing. Think about it: out of the billions and billions of snowflakes that fall, they are almost all based on the number six.

When I was a little kid and first learned that fact, I pictured little sixes falling out the sky, and covering the ground. I used to practice my multiples that way: 6, 12, 18, 24, 30, 36, 42, 48, 54, 60, 66, 72, 78, 84, 90, 96, 102, and so on.

One day when I was six, I got lost wandering around the neighborhood and my mom freaked out. She and my dad looked everywhere and finally found me under some bushes counting the multiples of snowflakes that trickled through the dense branches. Apparently, I was on the 778th multiple because I had counted to 4668. Looking back, I'm not even sure how I did that at that age, although numbers do speak to me sometimes. And NO, I'm not crazy.

So, you can imagine my shock and dismay when I took a careful look at one of the huge snowflakes and discovered that it was five-sided! I double-checked and wasn't imagining things. It was pentagonal with completely wrong angles. Gone were the beautiful 60-degree angles, replaced with the more obtuse 72-degree angles. It stopped me in my tracks.

"Matt, what's wrong?" Kelsie asked when she noticed I had stopped to look at the snowflakes on my fleece.

"The snowflakes …" I couldn't formulate adequate words.

"Yeah … what about them?" Kelsie urged.

"They're all … *wrong!*" I managed.

"*Wrong?* What do you mean?"

"They're five-sided!" Kelsie didn't seem phased, so after a while I explained to her about what is supposed to happen with snowflakes.

I found myself even telling her about the time I was under the bush counting by sixes. I'd never told anyone before.

She stopped abruptly. "What's going on? The clover. Ants with seven legs. And now the snowflakes."

"I take it you don't like my radiation theory." Although I was trying to joke around, I really was worried.

"No. And that still wouldn't explain the snowflakes."

"They're nuc-u-lar snowflakes," I said, and then immediately regretted it as I thought about nuclear winter.

"That's not really funny, Matt."

"Sorry," I said, catching up. "Well, was it a different species of clover? Were they different ... wait, are there different kinds of clover?"

"Yeah," she said. "Red clover and white clover are the main types. And then there's crimson clover and others. But that's not what I mean. I mean your numbers thing. What do you think it means numbers-wise?"

I laughed. "You know it could be dangerous asking me anything about numbers."

"I'm a bit of a thrill-seeker," she said sweeping her arms in a big circle around her, "shock me!" I wasn't sure if she was being sarcastic, but she seemed genuinely interested. And, of course, I can't resist talking about numbers.

"I ... I ..." I stammered. And for the first time that I could remember, I had nothing to say about numbers. Usually numbers, well, they speak to me in a way. They move around my head and arrange themselves into patterns that make sense, like puzzle pieces that self-organize. I can usually just describe what I "see." But right now I had nothing. It was like numbers were dead to me.

"What's wrong?" Kelsie stepped closer and sat down next to me.

"I'm not sure. I ... don't know what to say." For some reason, I couldn't look at her.

"Come on," she cheered, "do you remember that time in fifth grade when you worked out a formula for calculating the volume of a cone, or the time in seventh grade when you created a formula for estimating the circumference of an oval. Or the time in third grade when you explained the Pythagorean Theorem to Mr. Wolf and he stood there in front of the class with his mouth open like he'd seen a ghost."

I laughed, "How do you remember all ..."

"Wait!" she shoved her hand against my chest, stopping me mid-stride. "What was that?" Kelsie pivoted directly into a fighting stance, her hands balled into fists. She was hyper-focused and alert. I wouldn't mess with her, I can tell you that. "I saw something over there." She pointed to a small white tree. "Something moved over by that birch."

"What is it?" I asked, staying behind her.

"I don't know. I think ..." she trailed off as her head tilted to one side. "I didn't want to say anything before, but I keep seeing things out of the corners of my eyes."

"Peripheral vision?"

"Sort of. At camp they taught us something called wide-angle vision, which is good for seeing movement and ... there! Did you see that?"

"No." All I could see was a birch and some bushes. "What am I looking for?"

"Farther back," she said gazing blankly. "A shadow coming toward us." She tilted her head and paused for a moment, "I have a sense it's not harmful."

"Are you sure?"

"No."

"What do you mean, *no*? You said ..."

"I said I had a *sense* it's not dangerous. I'm not one hundred percent sure."

"Don't tell me that!"

"Okay, Matt. It's *not* dangerous and we have nothing to worry about. Just put your head in the ground like a good little ostrich." She took a step forward.

Now I'm used to sarcasm from most of my friends and I can take a lot of verbal abuse from people—hey, you have to in high school—

but Kelsie's little comment stung a lot. Let me repeat: a lot. I stepped up next to her and imitated her stance. I saw the corners of her mouth curl upwards into a slight smile.

So the two of us were standing there awaiting a possibly horrible death at the hands or paws or some vicious creature or maybe even a serial killer or something. We had gone through an abandoned tele-portal in search of a scientist who may or may not be alive, may or may not be hostile. And we weren't in Kansas anymore, Toto. Not that we lived in Kansas, but that's a reference to the 1939 *Wizard of Oz* movie, which I found as a fitting parallel to our current situation.

A shimmering gray form popped into view in front of the birch tree and floated toward us. It was approximately human size and shape and made up of an assortment of black and white squares and rectangles. It was partly translucent, and sort of looked like a Kindergartner's finger painting of a person. Something that caught me off guard, though, was that it had bright yellow eyes that made it pretty terrifying. I know Kelsie said she had a sense that it wasn't going to harm us, but let me remind you: IT WAS A GHOST! And it was coming toward us.

All I could think about was that this thing would come and suck our souls out of our bodies. Maybe I have played too much Dungeons & Dragons, or watched too many bad horror movies, but I thought of dozens of horrible ways this thing could kill us: drain our life force, freeze us with paralyzing touch, suck our youth from us and make us old, or maybe crawl inside our bodies and eat us from the inside.

Let me pause to talk about death for a moment, at least with math in mind. There are two main problems as I see it. One has to do with infinity and the other with zero. If people go to heaven or hell when they die, and either of those lasts forever, then we have to wrestle with the concept of infinity.

Forever. Never ending. Ever.

It's hard for the human mind to even try to comprehend such a thing. I mean, we can say we understand infinity, and we can write the symbol on a piece of paper. We can use the phrase from calculus "as X approaches infinity" to solve complex problems, but the reality is that if we try to imagine the universe as being infinitely large or atomic particles as breaking down into infinitely small particles, then there are serious issues. How do you picture such a thing? How do you wrap your mind around it? What box do you put it in? In other words, what does infinity look like?

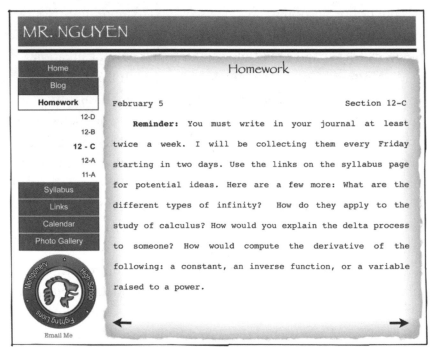

Every time you try to show me infinity, then there's still more. Try stopping anywhere to show me what infinity looks like and no matter where you stop, there's still more. And more. And more. I don't have a problem with heaven or hell existing, I have a problem with the mathematics of forever. I just don't know how to think about it. By the way, this is true of reincarnation as far as I can tell. The cycle of new lives could be infinitely long, so the same basic issue remains.

This might mess with your head. Which line segment contains more points?

Although it's tempting to think line segment EZ is the answer, the question itself is a trick. Believe it or not, they both contain an infinite number of points!

What becomes even more mind-boggling is that there are different types of infinity. For starters, there's positive infinity and negative infinity, which in my mind are really just infinity in different directions. Then there's *countable* infinity and *uncountable* infinity. An example of countable infinity is the set of integers. If you had an infinite amount of time you could count, well sort of, count to infinity—or I guess you could say that you could infinitely count out the integers, like: 1, 2, 3, 4, 5 … and so on. It would take you an infinitely long time, but in a manner of speaking, it could be done.

Uncountable infinity is different in that you cannot really count it out or get anywhere. The set of real numbers is an example. If you tried to count using all of the real numbers and you started at 1 and tried to progress toward 2, then you run into an immediate problem because the set of real numbers includes all of the decimals in between. So if you started counting, you might be tempted to go to 1.1, but first there's 1.01, 1.001, 1.0001, 1.00001 and so on. Wherever you want to start, there's always a smaller decimal. In fact, there's an infinite number of smaller decimals in between 1 and 2. And the true problem is that there is no place to start counting. That is why it is uncountable.

I like to think of these two types of infinity as horizontal and vertical. I picture countable infinity as a horizontal line, much like a standard number line. I picture uncountable infinity as a series of vertical lines disappearing down, like steps into infinity. Another way I think about them is that countable infinity is the infinity of addition (and subtraction), and that uncountable infinity is the infinity of division (and multiplication).

There are also different types of infinity in terms of how they appear. For example, there is infinity in one dimension, which is infinite length; in two dimensions, which is infinite area; in three dimensions, infinite volume; in four dimensions, which is ... actually, I'm not sure what that would be. Maybe infinite time.

What can get even more confusing is that there might even be an infinite number of infinities.

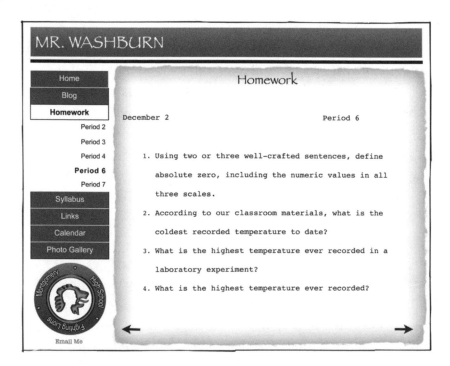

MR. WASHBURN

Home
Blog
Homework
Period 2
Period 3
Period 4
Period 6
Period 7
Syllabus
Links
Calendar
Photo Gallery

Email Me

Homework

December 2   Period 6

1. Using two or three well-crafted sentences, define absolute zero, including the numeric values in all three scales.

2. According to our classroom materials, what is the coldest recorded temperature to date?

3. What is the highest temperature ever recorded in a laboratory experiment?

4. What is the highest temperature ever recorded?

I know that you leave behind a body that will eventually deteriorate, feed worms, push up daisies and all that, but what about each person as an individual personality, or soul or spirit? Are our lives infinitely long? Do we live forever in heaven or hell? Do we reincarnate back to the planet an infinite number of times?

The other problem with death or dying, as I see it, has to do with the number zero. If people simply cease to exist when they die, if that's the end of everything in that life, then that's like zero. There's absolutely nothing left.

Nothing. Completely empty. Void.

Zero is bothersome, just like infinity. Actually, I should digress from my digression to clarify that zero and nothing are not equal. In other words, they're not the same thing. In a nutshell, zero is something that represents—symbolizes—nothing. Or it's nothing represented by something. Confused?

Let me give you an example. Show me zero apples. Now, you might be tempted to point to your open palm and claim that there are zero apples in your hand. But show me zero apples. In other words, show me the zero-ness of zero apples. You see the problem? It's not tangible. There's nothing to show. Besides, what you're pointing to is not zero apples, but actually air molecules and other things like dust motes.

Division by zero can really mess with your head. In basic mathematics, dividing by zero, has no meaning. If you were allowed to divide by zero you could really mess with the established rules of mathematics. For example, let's go through a little basic algebra. If you have an equation such as $2A = 6$, you can divide each side by two to isolate the variable "A" and arrive at the answer $A = 3$. As long as you divide each side by the same number, the equation stays equal.

If you could divide by zero, though, serious problems occur. Take, for example, the following equation: $0 \times 2 = 0 \times 3$. If we divide each side by zero here just as we divided by two in the other equation, then we get $0/0 \times 2 = 0/0 \times 3$, or simplified $2 = 3$. Can you see the problem?

If you divide by zero on some calculators you get an answer of infinity, and if the number you started with was negative, you get negative infinity.

The issue with division by zero can be thought of in these terms: How many times can zero fit into six, or how many times can zero apples fit into six boxes? The answer is an infinite number. In other words, you can always keep putting zero objects into six boxes because you're not putting anything inside.

What's even weirder is the thought of dividing zero by zero. It's different than dividing other numbers by zero. The question becomes: How many times can zero objects be put into zero boxes? The question itself is strange because in a way it's not asking for anything to occur. A lot of math books say that zero divided by zero is "undefined," or on my calculator, the quotient for zero divided by zero is "NOT A NUMBER."

So, with Kelsie and I standing there, I was wondering what would happen if this ghost thing attacked. I wondered if I would turn out to be "NOT A HUMAN" or if Kelsie would become "UNDEFINED." On this side of the teleportal, would our finite forms be infinitely divided, dispersing us into nothing at all? Would we transform from the finity of flesh into the infinity of spirit—or something like that?

Whatever the case, I was a little scared facing that ghost thing, mainly because I didn't have a sense of its intentions. Although Kelsie insisted that it didn't feel malevolent, I certainly wasn't sure—and I didn't want to find out the hard way. Kelsie stood her ground and I took courage from her determination.

As the thing moved closer, I understood what Kelsie meant by it being harmless. I know when I meet a stray dog on the street that I can immediately tell if it's dangerous or not. There's something about the way it approaches, the way it looks, the position of the tail and so on that tells me whether I can go pet it or not. Well, this was similar. Kelsie's arms dropped from an aggressive position down to her sides. I put mine down as well.

The next thing that happened was definitely the weirdest experience of my life, and I want to point out that I had already found a secret door in our school bathroom, decoded holographic puzzles, and I had gone through a strange portal to … well, somewhere. To say that this was the weirdest of all is actually saying something!

The ghost thing stopped in front of us. I watched as the black and white squares and rectangles shimmered randomly around its body, making it look like one of those basic composition notebooks. Then suddenly, a pattern held for about a second, and then it went back to being random again.

"What was that?" Kelsie asked.

"I'm not sure. It kind of looked like it just glitched or something."

We waited a bit to see what would happen and, once again, it was shimmering and random. Then, a pattern held for a brief moment.

"It kind of looked like a QR code," Kelsie said.

"Wait!" I practically screamed. "Kelsie, give me your iPhone!"

She took her phone out and handed it to me. I waited until the pattern returned and I took a picture. It actually was a QR code!

"Well?" Kelsie asked.

"It just communicated with us. It said, *I am Fifty-Seven. Please come with me.*"

"Fifty-Seven?" Kelsie responded. "What does that mean?"

I shrugged, "I don't know."

16

Meeting Maglio

The shimmering thing turned and ghosted away from us. We followed it for a while without saying much. I was still thinking a lot about my earlier observations and the differences in the things we were finding, not to mention dying, and Kelsie seemed focused on tracking where we were going and where we were coming from. On some weird level, the five-sided snowflakes still haunted me, perhaps even more than this ghost.

I kept taking pictures of it hoping it would keep communicating, but in its shimmering form, I could never get a QR code reading. I did officially name it Fifty-Seven, though.

"That's weird," Kelsie finally said. "I can't stop thinking about what it, um, communicated. What does it mean by Fifty-Seven? Is that how old it is?" Kelsie asked as she stared after it.

"I think that's its name: Fifty-Seven."

"But that's not a name," Kelsie said. "It's a number."

"Well, numbers are names for things."

"No, they're just numbers." She turned to look at me. "Wait! Don't look at me like that. I forgot I was talking to you."

"A number is a name for a quantity or a process."

"Quantity I get. What do you mean *a process*?"

"Well, there are several things, really. Numbers first have to be seen in relation to other numbers. For example, start counting."

We started walking again so that the ghost thing—I started thinking of it as Fifty-Seven—wouldn't get too far ahead of us. "What do you mean?" she asked.

"I mean just start counting."

"Okay … one, two, three …"

"See, you started with *one*. Why?"

"Because that's where you start."

"Why didn't you start at zero or negative three?"

"Because you don't do that. Everyone starts at one."

"Yes, but *why*?"

"I don't know. Why does the alphabet start with 'A'?"

I laughed. "Good question. I have wondered about that too. Let's just stick to numbers. One is *the* start of the counting numbers—at least in that direction."

"Direction?"

"Yeah, *up*. You started counting up instead of down. That's another part of the process. You know how when a rocket is about to blast off or the apple is dropping in New York for New Year's, we always count down."

"Yeah …"

"Well, one can be the last number, as well. Three, two, one, then there's the *Blast off!* or whatever. So one can either be the first or last of a process depending on which direction you're going."

"That's pretty cool. How do you think of all this stuff?"

"It's a scary world in my head, Kelsie. A scary world."

We were quiet for a while. I thought maybe she was bored with numbers or was focusing on tracking or something, but then she suddenly burst out. "Whoa!"

She said it so fast that I quickly looked around. "What?" For an instant I thought we were being ambushed or something.

"You know how you just said that one is the start?"

"Yeah."

"Well, switch the letters around in the word *one*."

"Eno?"

She laughed and tipped her head back. "No. *Neo*, as in the prefix for *new*. *One* means new." She paused and then looked up. "Hey, that was the character's name in *The Matrix*, wasn't it?"

"You're right! I never thought about that."

"The Matrix thing?"

"Either one. I never thought about one as a word, not really. But it makes sense. In either direction it can be seen as something new. In the case of counting up, it's the start to a new series, a new beginning and ..."

Kelsie jumped in, "And in the case of counting down, one is like an announcement of what's about to happen. Like the apple dropping. That's so cool! I didn't think numbers could be fun to think about."

I wish I had said something cool at that point, but I simply nodded.

"So what about two?" She was practically bouncing with excitement now. I had to admit I was pretty excited as well.

"Well ..."

"Wait! Two ... tow ... wot ... owt ... Hmm, *tow* is a word, but how does it make sense with what we're talking about?"

"Not sure. What do you think about when you think of the number two?"

"Twos. Pairs. Noah's Ark and all that. Opposites."

"Oh, maybe that's it. Two *tows* from one."

"What?" she asked.

"You just said *opposites*. Well, two is the second of the opposites. It gets *towed* from whatever one was."

"I'm supposed to understand that?"

"Pick something and I'll provide its opposite."

"You mean like night?" she said.

I nodded. "Day."

She pointed, "Up."

"Down."

"Left."

"Right."

"Forward," she huffed, sprinting forward a little.

"Backward," I said stepping back.

"Heaven."

"Hell."

"Female," she said pointing to herself.

"Male," I countered.

"Positive," she said pointing to herself. "Negative," she quipped pointing at me.

"Hey!"

We went on for some time, following the ghost thing and completing dozens of pairs that are apparent opposites. With my favorites being: yin/yang, good/evil, law/chaos, and dawn/dusk. It's a fun game, actually, because there are hard ones that really stretch your mind. For example: what's the opposite of a tree? What's the opposite of a lemur or Monday? The cool thinking part comes when you get difficult challenges like that. What happens is that you have to think about certain aspects of the word and then find an opposite. For example, Kelsie said *tree* at one point and I was stuck for a while until I thought about trees as the "lungs of the planet." Then I thought of lungs on a much smaller scale and finally said h*uman lungs*—which makes a certain kind of sense since trees and humans exchange oxygen and carbon dioxide.

The other cool thing is that every object has multiple opposites depending on *how* you think about them. I first thought of trees as lungs of the planet, but I could have thought about them as building materials. So, in a certain way, a tree could be thought of as an unfinished house. So *that* opposite might be a finished *house*. But you could also think of trees as the *oldest* living things in the forest, so an opposite might be the *youngest*, something like ferns or grass. Trees are often the *tallest*, so you could think about *shortest*, maybe microbes or something.

I can imagine a creativity exercise based on this to see how many opposites a person could develop for a single object or idea. In this case, a person would have to provide either the tow word (opposite) or perhaps the relationship. For example, if I said that whales and trees were opposites, then you would have to decide *how*.

You can test your thinking with the following example:

	Neo or First Word	Idea or Relationship	Tow Word (Opposite)
1		Lungs of the planet	Human lungs
2		Building materials	House
3		Oldest in the forest	Ferns
4	Tree	Tallest in the woods	Microbes
5		Tallest living things	Skyscrapers
6		Sunlight collectors	
7			Whales
8			Stones
9		Sources of fire or heat	

"You see, Kelsie? Every time you say something, I'm *towed* to an answer. I can't just pick anything."

"Oh, I get it. So if I said *tomorrow* you can't respond with *pineapple* or *cat* because you have to find a suitable opposite."

"Exactly, unless I can actually figure out a relationship. The second answer is towed by the first."

"Hmm. I guess. But it's not as cool as *neo*," she retorted.

"No," I laughed, "not as cool as *neo*."

The sun had gone down and it was starting to get dark, but I was too engrossed in our conversation and Kelsie didn't say anything about stopping to make camp, so we just kept following the ghost thing.

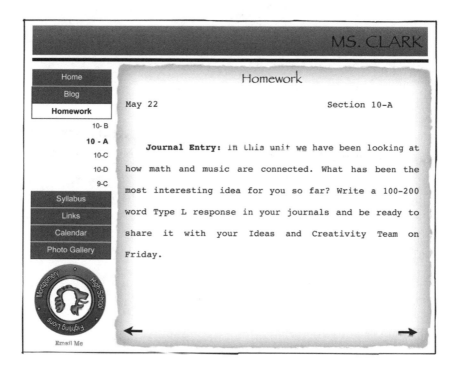

"What about three, Matt? Let's see, there's *the Re*, as in the Egyptian god, or *ether*, or *there*."

"Wait, how do you spell 'ray' as in the second musical note?"

"R-E, I think." I thought for a moment. "That's weird," I said, "so, three could somehow be connected with the second musical note? Hmm, maybe there's one missing."

"One what?" Kelsie turned.

"A musical note. Maybe there's a musical note missing and *Re* really is supposed to be third. Think about it: there's eight musical notes, right?"

"Well, seven, really, but *Do* is repeated."

"Well, either way, seven or eight, I think there's something missing."

"Missing?"

"Yeah, there are nine non-zero digits. I think a ninth note is hidden or something."

"How can a note be missing, Matt?"

"I don't know. Maybe it's like a dog whistle. The sound is there but we can't hear it."

"Then it wouldn't be a note, would it?"

I laughed, "I'm not sure. It's just a working theory."

"Or a not-working theory." Kelsie was about to say more, but then she stopped walking. "But why does there *have* to be? I mean, there's only seven colors in a rainbow."

"See, you're making my point for me! I think there are two colors missing in the rainbow! I think they are invisible or something … or maybe there's an infinite number!"

"You're weird, Matt."

"I know," I said, "but weird means I'm different and I like being different."

"You are definitely *different*," she said as she started walking again. We continued following Fifty-Seven.

I thought we were done with the conversation, but Kelsie started again. "Okay, what about *ether* or *there*? How do they make sense?"

"Well, there makes sense to me."

"*Why?*"

"Because three points can make a triangle, and triangles are the first shapes that can contain an area. In other words, the first time there can be a *there*." I pointed at the ground.

"Ah, that sort of makes sense. So, what about *ether*?"

I was about to offer an opinion when we crested a hill and I peered down at a curious scene. Down below, the ground leveled out and a circle of nine small shelters was arranged equidistantly around a central fire pit. The grass was worn into dirt all around and, although I couldn't see a fire, I could smell burning wood. Fifty-Seven floated down toward one of the far buildings, which I now noticed were simple thatched huts maybe six or seven feet in diameter.

The number nine caught my attention. A thought occurred to me that maybe each represented one of the nine Dungeons & Dragons alignments, which are like descriptors of how characters are supposed to behave. My next thought was of Dante's *Inferno*, where he described nine different circles of hell. Nine doesn't always do that for me, but maybe I was thinking too much about ghosts and wraiths and dead things.

Nine is a fascinating number, though. For starters, nine is the largest single digit, so it often represents or symbolizes completion. Groups of nine are called Enneads and there are a lot of famous ones. In Greek mythology, for example, there were nine Muses, or goddesses, who represented the nine arts:

- Calliope (epic poetry)
- Clio (history)
- Erato (lyric poetry)
- Euterpe (music)
- Melpomene (tragedy)
- Polyhymnia (sacred song)
- Terpsichore (dancing)
- Thalia (comedy)
- Urania (astronomy)

There were also nine Egyptian gods of Heliopolis, nine gods of the Etruscans and nine gods worshipped by the Sabines.

Nine is also used to describe personality types, generally known as the Enneagram. The nine I saw once were: *reformer, helper, achiever, individualist, investigator, loyalist, enthusiast, challenger* and *peacemaker*. In the Middle Ages, there was a similar idea called the Nine Worthies. These were nine historical figures whose combined qualities supposedly made up the perfect warrior: Joshua, David, Judas Maccabeus, King Arthur, Charlemagne, Godfrey of Boullon, Hector, Alexander the Great, and Julius Caesar.

Nine has a curious property that makes it very useful in a lot of those "magic" number tricks where you choose a number and after a few steps someone tells you the amazing result. For example, pick a number besides zero. Multiply it by nine. Add the digits together, and keep adding the results together until you reach a single digit. Result: nine. Try it again if you don't believe me. Just make sure to keep adding the digits together until there's just a single digit left. For example $9 \times 9 = 81$, and then $8 + 1 = 9$. Or 23×9 equals 207, and $2 + 0 + 7 = 9$. Here's another: $177 \times 9 = 1593$, and $1 + 5 + 9 + 3 = 18$ and then $1 + 8 = 9$ as a single digit.

This property of nine allows you to do something called "casting out the nines," which can be actually used to double-check arithmetic problems. Let me start with a small example using something called root numbering.

Take the number 96. If you add the digits together you get $9 + 6 = 15$. Root numbering involves adding digits together until there is only a single digit remaining, so the last step is $1 + 5 = 6$. So the root number of 96 is 6. But you could have arrived at the answer faster by simply "casting out" the nine, leaving the six. Or take the number 9689. $9 + 6 + 8 + 9 = 32$, and then $3 + 2 = 5$. But you could have just ignored the nines and arrived at the same answer with $6 + 8 = 14$, and $1 + 4 = 5$.

Because we're adding the digits together eventually anyway, you can cast out combinations of nine. So, for example, 75423 the long way is $7 + 5 + 4 + 2 + 3 = 21$ and $2 + 1 = 3$. But with casting out the nines we can match the 7 and the 2, and the 5 and the 4, which both add to nine, and cast them out. That just leaves the 3, which is the root number.

This property of nine, this personality that nine has, allows us to check answers in arithmetic problems. Here's an example using addition:

	4	3	2	8		=		8
+	1	4	4	0		=	+	9
	5	7	6	8				17
								1 + 7
Root: $5 + 7 + 6 + 8 = 26$ and $2 + 6 = 8$								8

Since the sum of the roots of the original numbers is 8 and the sum of the roots of the answer is 8, then we can conclude that we carried out the correct arithmetic. It's not perfect, since sometimes the root numbers can match and the answers might still be wrong because of some kind of error, but nothing's perfect. And it's fun, anyway.

The ghost-thing, Fifty-Seven, changed direction slightly and shimmered over to the far side of the fire pit, where it stopped. Another pattern formed temporarily, and I snapped a quick QR photo. Kelsie turned to me, "What did it say now?"

"It just said, *They're here.*"

I heard a noise down by the fire and a figure which had been blocked by one of the huts stood up and turned to face us. A man's voice called out, "Well, you are not Doctor Morris!"

It wasn't much of a hello, but I was happy to see another human being. Kelsie stepped down the hill and I followed. As we got closer, I could see that the figure was thin and about as tall as me. He stepped toward us and I saw he was a middle-aged man, maybe in his early forties. He was dressed in a ragged dress suit. He had a piece of blue cloth tied around his head keeping his wild hair from spilling into his face. His cheekbones were a bit sunken and he had dark circles under his eyes. Fifty-Seven floated to a stop behind him.

Kelsie stepped forward. "You're Maglio, aren't you?"

The figure stared back at us. "Who, who, who, who, who, who, who, who are you?"

Kelsie and I looked at each other. "You are Maglio, aren't you?" Kelsie repeated. "I, uh, *read* your journal entries. It is you, isn't it?"

The figure nodded. "Yes, I am, um, Maglio Mordavarian, bumbling scientist, lost explorer and probably eight other things as well. And you are my help?" Kelsie and I looked at each other again, but didn't say anything. "What kind of help are you anyway? I expected about eight or more a lot sooner."

He stepped closer, "Where are ..." He stopped and looked us up and down very carefully. "You are just kids! They sent graduate students to help on something this important? Unbelievable!" With that he turned his back on us, walked to his spot and sat down on the log where we first saw him. Fifty-Seven followed closely behind.

Kelsie and I edged over closer to the fire. The heat felt good. I could see that Maglio was leaning forward and rubbing his forehead. "Where are the new instruments and the ..." he looked up again. I had no idea what to say. Kelsie opened her mouth to say something, but Maglio cut her off. "You're not even graduate students are you? And by the looks of things, you didn't bring much equipment with you. So who are you and how did you get here?"

Kelsie and I slid our backpacks off and sat down near the fire.

Kelsie spoke. "Mr. Maglio, we're ... um, this is Matt and I'm Kelsie. We found your ... Matt found the sink computer in the school and ..."

"School?" Maglio frowned.

"Yeah, our building is being renovated, so we're in a different ..."

"What happened to AREA?"

"I ..." she looked at me, "well, we don't know about any of that. Matt found a computer in the sink and ..."

Maglio smiled. "Ah yes, that was my idea, you know. The sleek, hidden design, water safety features, the mirror display, and every-thing."

Kelsie continued through his interruption, "... and Matt made a plan to break the code so that ..."

"I knew I should have encrypted that further. Budget. Everything is based on a budget these days." He stood up and paced around in tight circles. After a few rounds he suddenly sat back down. "And, of course, I did not really want ... oh, never mind."

"Um, so anyway, we broke the code," Kelsie paused to look at him carefully. Maglio was still frowning from the budget comment. "Then we found the lab and the journal and began to piece together what happened. We got everything working and …"

"How did you do all of this?" He sounded remarkably doubtful.

"It was Matt, mostly, he and Ari broke the codes and got into your journals. John wired the teleportal to work. And Thomas and Jamie, well, they mostly listened to your albums."

"What did you think of my three-dimensional turntable?"

I finally jumped into the conversation, "That thing is so cool! How does it work?"

Maglio had a big smile on his face. "It is basically just mathematical translations of laser images of the album's surface. Some of my best work, actually. Now, I assume you used the emoti-promptor?"

"The stylus-drawing compass thing?" I asked.

"Yeah, what did you think of that?"

Kelsie jumped in, "I love it! I always loved reading, but this is completely different! It's like you're really there inside the reading, experiencing everything."

"Nano-feedback isn't exactly my specialty, but there were just too many practical applications."

"I know! This kind of reading—we started calling it dyna-reading—could change everything! People could read and experience entire libraries!"

"Yes, I should think that the promptor alone would be worth a Nobel Prize or eight, but Maglio digresses."

He turned to me. "Now, how did you unlock the journal screens? No, never mind. You did and you are here, which means that unless this is a powerfully humorous joke—ha, ha, ha, ha, ha, ha, ha, ha—then you're telling me the truth. Wait? How did you figure out how to activate the teleportal, anyway? I was rather proud of that one. By the way, can you please say the number 8 eight times each."

"Um …" was all I could manage.

Maglio leaned forward and looked at us at a weird angle.

Kelsie and I looked at each other and then started together, "Eight, eight, eight, eight, eight, eight, eight, eight."

Maglio looked around and then shrugged. "Okay, continue."

I explained to him just about everything that we had gone through. He listened carefully and smiled a lot whenever I spoke about one was of his inventions.

"You know, I went back and forth about all of those codes. Part of me wanted to secure everything with serious encryption software, and another part of me wanted to use time-secured vaults and such. Another part wanted to make everything public and there was even a part that wanted to destroy everything. But something nagged at me the whole time to keep the work I had done partly accessible, at least to the right people. I just never imagined it was going to be college students." He got up from his seat and started pacing. Fifty-Seven followed him.

Maglio continued, "Heck, I don't think Morris could have even cracked the start-up code. He's too fixated on exact details and would never have thought about pi in its more beautiful and symbolic form." He spaced out for a moment. "You know," he winked, "the definite is better than the indefinite."

I nodded. What else was I supposed to do?

"Hmm, well, anyway, I am glad I listened to my instincts. Otherwise, the two of you would not even be here. But there are a lot of questions. A lot of questions—at least eight."

He turned to the ghost, "Fifty-Seven, what are we going to do?" It didn't move and if it acknowledged what Maglio said, I couldn't tell.

Kelsie jumped in, "Why Fifty-Seven?"

"Because I love ketchup."

17

Eight Wonders

With Maglio's responses—strangely filled with eights—I have to say I wondered about his sanity. In all of the deserted island movies I've seen and books I've read, it doesn't take much to tip someone over the edge from normal to *Because I love ketchup*. What kind of answer was that?

After his ketchup response, Maglio turned to look at Fifty-Seven and smiled. He sighed heavily and then sat back down. I realized then just how tired I was. My feet hurt. My legs hurt. In fact, most of my body hurt. I was sore, tired, hungry, and I have to say even remarkably grumpy, despite spending so much time with Kelsie. I had no idea how long we had been in this strange place. Hours certainly, but I really wasn't sure. I could tell it was getting really late, though, because I could barely stay awake.

"There are things I have to tell you about this place," Maglio said. "I am sure smart kids like you have already figured out that this is not Earth—at least, not the Earth to which we are accustomed …" He trailed off for a bit, "But you are obviously tired and …"

"Wait," Kelsie said, "one thing first. Are we going to be able to get back?"

Maglio stretched out his arm and tapped a large wristwatch. "Yes. Now that the portal is activated on the other side, we can go back once I recalibrate the one on this side. Thank you very, very, very, very, very, very, very, very much for reopening it, by the way."

Kelsie and I both nodded.

"However, it is getting late, and Fifty-Seven and I have, um, *eight things* to do tonight. Let us call it a day and I will explain what I can tomorrow."

The way he said *eight things* was creepy and unsettling, but I was too tired to follow my instincts and just let it go.

Tired, that is, until I heard the next words out of Maglio's mouth. "Come on. You two can sleep over here in hut eight." With that he motioned toward one of the small huts. I have to say that I almost choked and I'm sure I turned a bright red. Maglio must have noticed my sudden embarrassment, because he quickly amended what he said. "Actually, you can each have one. Just make sure one of you is here," he said pointing to the one I had associated with Terpsichore (dancing) and the personality type of *challenger*. "It's, uh, the best one here."

I deferred to Kelsie, but she just shrugged. "I can take that one." She looked at me briefly and then fished into her backpack. "Here, Matt, I brought a sleeping bag for you. And a ski hat. If it gets cold, just put it on. And if it gets really cold, you can … um, never mind." She broke eye contact and looked up suddenly, which I was happy about because I felt my face flush. She continued, "I don't think it will get too cold."

"Thanks," I said, turning away. I took the sleeping bag and the hat, and walked to one of the other huts. I turned around to check on Kelsie. She crawled into her hut and Maglio seemed headed for what I thought of as Melpomene (tragedy) and *achiever*.

I pulled open the small thatch door and sleepily clambered into the hut I had associated with Thalia (comedy) and *enthusiast*. Inside I found a small mat made out of pine branches. Not the most comfortable thing in the world—or in *a* world—but I was so tired that I could have fallen asleep standing up.

I wish I had asked Maglio why there were nine huts. He didn't mention any other people around, and as this thought slowly meandered its way through my head, I fell asleep.

The next morning when I awoke, I sat up and realized that a number of spots around my body were sore from being jabbed with pine branches. I don't know how people go camping without air mattresses

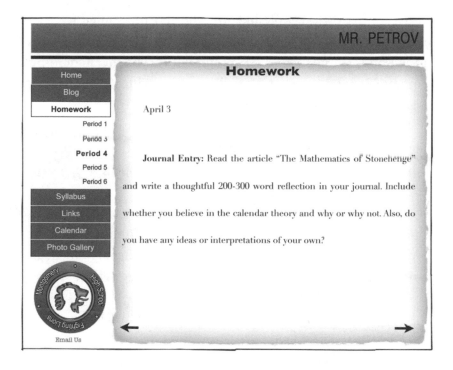

MR. PETROV

and actually enjoy themselves. Kelsie goes camping all the time and she even makes her own shelters at wilderness camp. I'll admit that having the skills to build shelters, collect food, make fires and hunt animals is really cool. But here's my question: Why? We live in a time of puffy mattresses, Spam, portable heaters and let's not forget smart phones. Why would anyone ever want to go back to prehistoric times? Or even pre-electricity?

She told me that one time she had to make her shelter, find her own dinner and even start her own fire. And she did all of it with only a knife—that is, if you count grass seeds, berries, and weird leafy vegetables as dinner. Where is the fun in that? I have to admit that making fire by what Kelsie called *hand-drill*, is pretty cool, though. There is something primal and mysterious about fire.

Kelsie said she likes that she can take care of herself. If something catastrophic happened, then she could go off into the woods and survive on her own. People like me who rely on canned food and microwaved meals would die if we had to spend even a few nights out in the woods. I understand her point, but maybe there could be a compromise, for me at least. How about a cave with a built-in pantry, an air mattress and satellite TV?

I crawled out of my hut and saw that the sun was already up and the sky was clear. Kelsie was standing at the fire pit, poking the coals with a stick. She turned when she heard my door open. "Hey, sleepy."

"Good morning. How long have you been up?"

"Maglio and I were up a while ago. I don't know, maybe two hours. I couldn't sleep in that hut, so I came out to sleep by the fire." She smiled, "Maglio woke me up this morning talking to himself." She poked the coals again as I stood up and walked over to her.

"Matt, do you think he's okay?"

"Maglio? I don't know, I mean, he seems a bit, you know …"

"Crazy?" she suggested.

"Yeah. I was going to say *nuts*, but I'm too hungry to think about food."

"You just said it, though."

"Shh, don't tell my stomach," I said imitating Corey Meunch's voice. She snorted and shook her head. There was something wonderful about watching her laugh.

"Matt, I know you love math and all, but even you don't talk to yourself in numeric codes."

"13-1-25-2-5, 9, 4-15."

"Matt!"

"Maybe I do," I said.

"See? How do you do that? How do you keep all those numbers organized?"

"That's just a simple number and letter cipher. Anyone can do that."

She frowned. "No! Not everyone can do that."

"Well, it's relatively easy, you just number the letters of the alphabet. So, A is one, B is two and so on."

She looked mad. "I know *that*. But they just don't stay put in my head. It's like they don't behave or something."

"Yeah, but words do that for you. You get a 100 on every vocabulary quiz. Well, now that I think about it, you get a 100 on just about every test you take."

She exaggerated a huge smile. "What I really want to know is: how do you and Ari play chess together without a board?"

"I don't know, Kelsie. We just can. It wasn't like my parents made me do chess meditations or sent me to chess camp, or anything. I don't think Ari knows either, although I heard a rumor that his parents beat him with a chess board."

"That's not really funny, you know."

"Sorry," I said. I pulled up a small chunk of wood and sat down next to her. I would have preferred a nicer chair, but it was better than sitting directly on the ground. "So, what do think about Maglio?"

Kelsie was quiet for a while as she poked the coals in the fire. "Um, the word *eccentric* comes to mind."

I nodded. "He keeps saying stuff about eight."

"But why *eight*?"

"I don't know, know, know, know, know, know, know, know."

"Matt!"

"You need seven more Matts."

Kelsie punched me.

Eight times.

I was about to say "Ow" eight times, but it seemed like too much. Plus, she gave me one of those *I'm going to kill you* looks. After a while she said, "You know, I couldn't tell if Maglio was happy or not to see us."

"What do you mean?"

"Well, I expected him to be happier that we were here."

"I know, but he's distracted or something."

"Well, he has been here a while."

"Still. It's like he doesn't care that he's been stuck here or something."

"But you heard him: we can go back through the portal now." I hesitated, but then put my hand on her shoulder. "We'll finish rescuing him today and then we can work out what to do about being grounded. And detention." I tried to say everything as a joke, but Kelsie didn't laugh. She did, however, reach across with her other hand and put it on mine.

She shrugged, "My parents are going to ground me until my wedding—if they ever let me out of the house to meet someone worth marrying."

I was about to respond when Maglio came over the small hill. "Good morning! Good morning. Good morning, morning, morning, morning, morning, morning. I'll bet you two have a lot of questions." Fifty-Seven was nowhere to be seen.

Kelsie and I looked at each other.

"Before I begin, though, how about some breakfast?" He reached into a small pouch at his waist that he had made out of a t-shirt and held out a handful of berries. "Go ahead, have eight or more."

Kelsie stood up. "Berries? Where did you get berries? I've been keeping my eyes open for food and haven't seen much except acorns." Maglio walked up to us and dropped a small batch of berries into Kelsie's hands. Then he dropped some into mine. Some breakfast. They were good, especially after a bland powerbar, I have to admit, but not exactly bacon and eggs with toast.

"Eyew, acorns are horrible," Maglio made a sour face, "I do not know how squirrels eat more than eight."

"Did you boil them first?" she asked in between stuffing berries into her mouth.

"The squirrels?"

"No! The acorns. You have to boil them multiple times and then they taste better. You have to get out the tannic acid."

Maglio swept his hand in front of him. "My kitchen is being remodeled and I am busy with eight things. I cannot boil much right now."

Kelsie brightened, "We can boil some water in my coffee can." She looked around the camp, "and I could get more going. We'd need some rocks—not river rocks—and a hollowed out stump and …" She started bouncing around again.

"Really? Do you have any tea bags by any chance?" Maglio asked.

"I could harvest some white pine needles, some mint and maybe wintergreen and …"

Maglio put his hands up, interrupting her yet again. "Hold on, let us do that in good time. I have some questions for you and then I am sure you have at least eight questions for me. Plus, there are some complications."

I shifted some berries in my mouth so I could speak, "What do you mean *complications*?" Some juice dribbled down my chin and I wiped it on my sleeve. Then, in a panic, I remembered it was Kelsie's shirt and looked over to see if she noticed. She hadn't. I checked the sleeve and luckily the juice sort of matched the color and wouldn't create a noticeable stain.

"Hold on. There are some things that first need answers. Last night you two said that your college is renting …"

"Uh, actually it's a high school," I said.

Maglio turned and looked at me. His face was as blank as an undone worksheet. If he was upset, he didn't show it. If he was happy, he didn't show that either. "I thought you two looked young. So, you two are in high school? And you did all this?" He swept his arm around in big circles, "All, all, all, all, all, all, all, all of this?"

"Yeah," I said. "We just got carried away with everything. It was better than most of our classes—except for math, of course."

Kelsie rolled her eyes.

"You two want jobs guaranteed for eight years when we get back?" He laughed and popped a few more berries into his mouth. "So, your high school is renting that space. 3-9-1 Merchant Avenue? Three story brick building? Double glass doors? Eight windows along the south side?"

"Yeah," Kelsie and I said at the same time. I have to pause a moment just to point out that 391 in the simple alphabet cipher is C-I-A. Probably just a weird coincidence, but still very cool.

"Think back a while. What was the building before you rented it?"

"It was empty," Kelsie said. "I drove by it every day to go to school."

"Empty? For how long?"

"I don't know." She turned to me. "What do you think, Matt, two or three years?"

I shrugged, "That sounds about right."

Maglio just nodded. "Less than eight, though? Okay, so you found the lab, opened the journals, and reactivated the teleportal. Then the two of you came through?"

"Yeah," we said again in unison.

"So who knows you are here, exactly?"

Kelsie and I looked at each other. She answered, "Well, no one was with us. But the others will figure it out when we don't turn up."

"Are they likely to come through the teleportal?"

"We're not sure," Kelsie said, "Matt and I finally figured out the whole *elements of caution* thing and I sort of dragged him in."

"So the others are unlikely to follow?" We both nodded. Maglio calmly pulled his bandana off and set it on his lap. He pulled his hair back with his hands and shook it wildly. Then he pulled it back again, tipped his head backwards so that his hair would stay and retied the bandana. With his shoulders back, his suit coat opened and I noticed that he had a pristine gray t-shirt on underneath. There were no tears, no stains, not even any ragged edges.

I must have been staring because Maglio looked down at his shirt and then plucked it with his fingers. "Nano-fibers," he said simply. "One of my other nanotech designs. They are self-cleaning and self-repairing." He wiggled his body, "Plus, moving around generates enough electricity to power a small device."

I laughed, "That is so cool! So you could plug a phone or MP3 player into your t-shirt!"

"Well, yes. Although I have another t-shirt design that actually is an MP3 player." He grinned, "Better overall sound and no need for earplugs." I had to admit the concept of digital clothing was way cool, but a bit terrifying as well. Kelsie didn't seem nearly as excited.

"Anyway, as you likely gathered, I was commissioned to work on a teleportation device. The military was thinking of using it for pre-emptive strikes, military coups, and eight or more other such belligerent ideas."

"And let me guess. Something went horribly wrong, just like it did in *The Fly*," I said.

"Which version did you see?" Maglio asked.

"The one with Jeff Goldblum," I answered.

"I like the nineteen fifty-*eight* version with Vincent Price better."

"At least we're not turning into flies," Kelsie laughed weakly.

After an uncomfortable lull, I broke the silence: "So, where are we exactly?"

"Ah, now that is the question," Maglio said simply. "I don't know."

"You don't know!" Kelsie blurted.

"I don't know *where* exactly, but I can explain some things. The teleportal is based on principles from quantum mechanics. Have they taught you that in school yet?"

Kelsie and I nodded. "A little," she said.

"Okay. So, according to some theorists there are multiple universes that coexist. Most agree that there are likely more than eight. Look at that tree over there. In some parallel universe that might be a maple, not an oak."

"Like Michael Crichton's *Timeline*," I said. "That was such a cool book! But they sort of went back in time."

Maglio nodded. "So, we seem to still be on Earth, but the real question we should be asking is: *When*?"

Kelsie jumped in. "If this is Earth, why are so many things different? Why does all the clover have four leaves? Why do the ants have seven legs ..."

"The ants have seven legs?" Maglio asked.

"Yeah," Kelsie continued, "they all seem to have seven legs."

"Hmm. Asymmetrical leg structure? That is rather interesting ..."

I jumped in, "And the snowflakes are five-sided!"

"Well, yes, I did notice that a while back ..."

Kelsie piped in again, "What I want to know is: where are the other animals? I've hardly seen anything higher in the food chain, like ... I don't know ... a hawk or something."

Maglio frowned, "They are around but in different *numbers*."

Something about the way he said the word numbers made me sit up. An image of Fifty-Seven popped into my head. That's when a lot of things suddenly made sense to me.

"Fifty-Seven is a number," I blurted out.

"Wow! Very good, Matt." Kelsie started fake clapping.

"No, no." This time I stressed everything, "*Fifty-Seven ... is ... a ... number.*"

She stopped and looked at Maglio, "What does he mean?"

"I was coming to that," he said. "Eight, for example ... um, yes, Fifty-Seven is a number. A *living* thing. *The* number 57. It is as alive as you and me."

"Alive?" she muttered.

"Yes," Maglio said. "Here, wherever, whenever we are, numbers are alive, *the* numbers are alive. Eight, for example." He looked around and I followed his gaze, but didn't see anything unusual.

"Where are all the others?" Kelsie asked.

Maglio sighed. "That is a part of the bigger picture. Fifty-Seven is searching for more as we speak. Hopefully, he will be back soon."

"*He*?" Kelsie said.

"Yes, odd numbers are male. Even numbers are female," Maglio stated.

Let me pause to explain a little here. In traditional numerology what Maglio was saying was true. The way the ancient Greeks thought about numbers meant that odd numbers were masculine and even numbers were feminine. As far as I know, the number one wasn't really considered even or odd. It was like the creator of all other numbers because it can't be split. In other words, it only has one factor: itself. Other numbers, even prime numbers, have at least two factors.

All through school I wondered what made odd numbers odd. Were they strange? Were they problematic somehow? When I started reading about Pythagoras, Euclid and other mathematicians, I began to understand more. Odd numbers can be split into three parts, two of them paired, with the number 1 always left alone. Take seven, for example. It splits into three matched pairs, with one leftover: 2-2-2-1. Or it can be a pair of threes with a leftover, such as: 3-1-3 (my favorite number). Nineteen, for example, can be split into nine matched pairs and one leftover: 2-2-2-2-2-2-2-2-2-1. It could also be a pair of nines and a leftover, such as: 9-1-9. With odd numbers there's always a leftover.

Think of the expression *Two's company, but three's a crowd*. There's a natural pairing and then one leftover. The single number that remains is like a single person; it just hasn't met another number yet. It's not complete in the sense that there's always an opposite complement awaiting a match. Think about opposites again: up/down, left/right, male/female, day/night, light/darkness. There can't be one without the other. For example, if there were only darkness, then the concept of light wouldn't exist. As soon as light exists, then there's a pair. Think about it: this is why adding two odd numbers make an even number and the sum of two even numbers make another even number.

By the way, this doesn't make men odd.

According to the ancient Greeks, even numbers are feminine. They are always divisible into equal pairs, with nothing left over. In other words, they split into even pairs, like six as two pairs of three, or eighteen as two pairs of nine. Even numbers, because of their completeness, are supposedly harmonious and calm.

All of this made a strange bit of sense when I took chemistry two years ago. Molecules or atoms that have a net charge are called ions and they are highly reactive because, basically, they haven't bonded with something else yet. Neutral atoms or molecules, on the other hand, are essentially non-reactive. So in this way of thinking, odd numbers are ionic and even numbers are neutral.

"Maglio, why fifty-seven?" Kelsie asked.

"I already told you," he smiled, "I *love* ketchup. Love times eight ketchup."

18
Tests

Maglio continued speaking, "There are a few problems that you two should know about in case Fifty-Seven returns with any additional numbers." Maglio stood and then paced in a figure eight. What was with all of the eights? Maglio carried on, "The first is that all of the numbers have different personalities and ..."

"Wait," Kelsie interjected, "I heard you when you said that the numbers were alive and I didn't say anything. Then you say they're male and female. I'm still not exactly sure what you mean by that, by the way, but now you're telling me that numbers have personalities? What are you going to say next? That they're dangerous or something?"

Maglio looked very serious. "Actually, that was not next, but if you want to talk about that, we can. Some of the numbers do not ... appreciate me being here, and I certainly think they will not be happy that three of us are here now. Some of them have tolerated my presence, but they haven't done much because I am just one human. Three humans? That might be too many." He continued pacing in a figure eight.

"We're the only people here?" Kelsie inquired.

"Yes. There *were* more of us here," Maglio sighed, "but they, uh, went back through to resupply and I never saw them again."

"The teleportal was burned out on our side," I said. "Could that have something to do with it?"

Maglio closed his eyes, lowered his head and took a deep breath. Then he suddenly stood up and turned his back on us. He exhaled loudly. "I was afraid of that," he said slowly. "My colleagues are likely ... lost."

There was a long period of uncomfortable silence. I kept looking at Kelsie and she kept looking back at me. We made various faces and mouthed words back and forth:

Say something!

No, you say something!

Go over there!

No way. You do it.

I don't know what to say.

Neither do I.

Finally, we just sat there watching the fire. Maglio got down on one knee and then made a series of gestures with his hand. I couldn't be sure, since his back was turned to us, but it looked like he was crossing himself repeatedly.

He eventually stood up and turned around. "Okay, they taught you about the first law of thermodynamics in school, right?"

"Yeah," I said.

"Okay, it states that energy cannot be created or destroyed, right?"

"Yeah …" I added.

"So what happens when you think about something? There is energy involved, right?"

"Yeah," Kelsie sounded interested.

"So, what happens to the energy of thought? What happens when I think about, say, the number Fifty-Seven? What happens to it?"

I think I knew where he was going and ventured a guess. "It collects somewhere. In this case, *here* in the form of what Kelsie and I have been calling a ghost."

"Well, Fifty-Seven is a ghost in a certain way of thinking. He is not physical in the same sense as you or me or that tree over there. But it does have a form. Whenever someone thinks about the number 57, then Fifty-Seven gets a little energy, and that energy takes the form of a cloud of potential energy."

"So, what you're saying is that when we think about something, it gets powered? All that power of positive thinking stuff actually works?" Kelsie asked.

Maglio shrugged, "Well, basically, yes."

Kelsie turned to me. "Matt ..." was all she said.

"I don't know what to say, Kelsie. This is weird to me, too."

"Yes, but not as weird. You already think about numbers all the time and ..."

Maglio suddenly snapped. "What did you just say?"

"I said: *Matt already thinks about numbers all the time.*"

Maglio turned to me sharply, a look of something between fear and wonder, like the looks that people give Corey Meunch when he eats something disgusting. "Is that true, Matt? You think about numbers all the time?"

He was so intense that I wasn't sure what to say. "Um, yeah, sort of."

"You either do or do not, which is it?" He was insistent. And I have to add a bit scary and intense, especially as he approached.

"Yeah. Yes, I do. It's just a hobby, or ..."

Maglio turned to Kelsie, "And you?" He turned to Kelsie. "Do you think of numbers all the time too?"

"Uh, not so much," she shrugged.

Maglio turned back to me in a snap. "Matt, what is your favorite number?" He was right in front of me now, and I have to say not only did he need a breath mint, but a bar of soap would have been good too. And some deodorant. And toss in some body spray.

"What?" I squeaked.

"I said: *What is your favorite number?*"

"Three hundred thirteen."

"A three-digit number! Interesting! Interesting!" He was almost yelling now. He backed off and began circling again, only this time in smaller, tighter circles. Then he turned to Kelsie. "And yours?"

"Sixty-one."

He stopped. "Sixty-one? Also interesting. Any reason?"

I shook my head to steer him away from the topic of Kelsie's dad, but he didn't seem to notice. When Kelsie didn't answer, he stopped circling. "Well? Why sixty-one?"

Kelsie looked down to where she was digging her foot into the ground. "It's, uh, my dad," she said softly.

Maglio looked back and forth between us. "Your dad, huh?" I could tell he was thinking deeply about something because his face was all scrunched up. "You two are interesting, that is for sure. This could work. If we just …" He turned and walked away, trailing off into a string of numbers. I only caught some of what he said. There was something about "huts" and "circles" and "dangerous," but I wasn't able to catch a whole lot more because Kelsie was suddenly next to me.

"See what I mean?" she whispered, "He's a bit … unsteady to say the least."

"Let's call him Crazy Eights."

"Matt! That's mean. He's just, you know, losing it a bit."

"That will be me if we don't get more to eat," I said. "Berries? I can't live on berries."

"Actually, you can survive on berries and a few other …"

"I said *live*, not survive."

She laughed a little and then was quiet for some time. "Matt, what's going on in that 'lil 'ole head of yours?"

Before I could answer, Maglio turned suddenly and snapped, "Come along, Matt. Time for a few tests." Maglio walked past the small hut that I associated with Urania (astronomy) and the *investigator* personality. Kelsie and I followed him approximately forty-seven yards to a flat patch of dirt that he had apparently been using as a kind of chalkboard. Forty-seven, by the way, is a cool number that shows up repeatedly in episodes of *Star Trek: The Next Generation* and the other spin-offs.

Maglio continued. "Matt, I am going to need to test you before I can finalize my … er, before we can figure out what to do next."

"What tests … exactly?"

"Just some mental problems at first and then I will get progressively harder until you get something wrong or you can't answer."

That didn't sound good. It sounded like he was going to keep asking me questions until I failed. Not good.

"Okay, Matt. Let us start with something easy. How about sixty-four times one hundred twenty-five." He snapped his fingers.

The game was on.

The thing about math, the beautiful thing about math, is that it follows rules. *Always* follows certain rules, actually. Numbers and equations can be rearranged as much as you want as long as you stick to the rules. For example, with multiplying two numbers, if you double one of them, and halve the other, then the answer will remain the same. For example, 4×8 (32) yields the same answer as 2×16 (32) and 1×32 (32).

Sometimes a rearrangement is easier to solve than the original problem, mainly because certain operations are easier to perform mentally. For example, in England one of the tax rates is 17.5 percent. This might at first seem like a really weird and awkward number to work with, but a closer look makes it relatively easy to calculate. 17.5 percent can be rearranged as 10% + 5% + 2.5%. Ten percent is just a decimal place shift; five percent is half of that and two point five percent is half again. So, 17.5 percent of 400 is just 40 (decimal shift) + 20 (half) + 10 (half of that), or 70.

So a trick for mental math is to find ways to rearrange problems to make them easier to calculate. In this case, I thought about a few different arrangements, but settled on:

$64 \times 125 =$
$32 \times 250 =$
$16 \times 500 =$
$8 \times 1000 = 8000$

"Eight thousand," I answered after a second or two.

"Not bad. Not bad," Maglio responded with a strange, intense look in his eye.

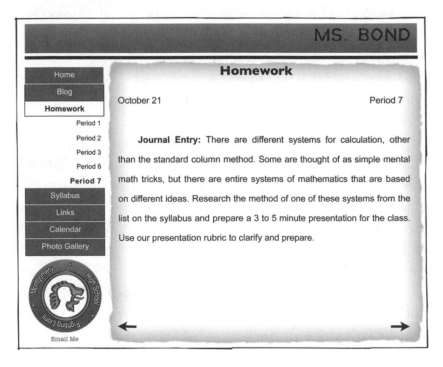

"How about seventy-nine times eighty-two." He snapped his fingers again.

This problem might be a little tricky for people to do mentally using the standard column system for multiplication. There are the initial multiplications, numbers to carry and then the two rows that must be added together at the end. However, there are other systems for math that may be easier. One of the systems that I like is from Vedic mathematics, which was developed in India early in the 20th Century. There's another one called the Trachtenberg system that's pretty cool too. If you're having trouble in math, it may help to try out some of these other systems. Maybe one of them will work better for the way you think about things.

In the Vedic system, 79 × 82 is split up and calculated "vertically and crosswise." So the first step is to arrange the numbers as:

7   9

8   2

Then multiply vertically and diagonally. The two diagonals get added together, so keep them separated at first. This is what the next step looks like:

$$7 \times 9$$
$$8 \qquad 2$$

$$(7 \times 2) + (8 \times 9)$$

56	(14 + 72)	18
56	**86**	18

There are two carries involved. The bold "8" is actually in the hundreds place, as is the 6 in 56, so it carries. The bold "1" is actually in the tens place, so it adds on to the 6 in 86 to make:

$$
\begin{array}{r}
5\,6 \\
8\,6 \\
1\,8 \\
\hline
64 \quad 7\,8 \;=\; 6478
\end{array}
$$

"Six thousand four hundred seventy-eight," I answered quickly.

193

Maglio paced in figure eights around me. "Pretty good, Matt. Mental arithmetic is there. Now, let us see if you understand mathematical patterns. What is the answer to this?" He picked up a stick and scratched tick marks in the dirt with a multiplication symbol:

$$111,111,111 \times 111,111,111 =$$

I certainly didn't know any mental math tricks, but there is a pattern involved, and math is often about finding patterns. When you multiply any number by one, the answer is always that number. That's known as the Identity Property because the other number keeps its own identity. With a problem like this, it's easy to get intimidated or overwhelmed, so always start small and search for a pattern. Here are a few smaller examples.

Take 11×11 first. In the normal column method, we would do the following:

$$
\begin{array}{r}
11 \\
\times \quad 11 \\
\hline
11 \\
+ \quad 110 \\
\hline
121
\end{array}
$$

Notice that it's really just two elevens stacked on each other, but shifted over one place value in order to deal with the tens place. So we get 121 as an answer because two of the ones (in the tens place) are added together in the middle, but the outer ones are by themselves. Watch what happens with 1111×1111:

				1	1	1	1
			×	1	1	1	1
				1	1	1	1
			1	1	1	1	0
		1	1	1	1	0	0
+	1	1	1	1	0	0	0
	1	2	3	4	3	2	1

Here we have four rows stacked and shifted for place value. It's the shifting pattern that's important. Moving left to right or right to left, each column gains or loses a 1, basically creating a counting pattern.

So in this case there is a long series of ones, nine digits stacked on top of each other in the standard column format. This means there would be nine columns in the answer and only the middle column would contain all nine ones. The other columns both to the right and left are shifted, so they contain one less 1. The answer then, would essentially be counting 1 to 9 and back down to 1, or: 12,345,678,987,654,321.

The harder part of this one was actually naming the number. "Twelve quadrillion, three hundred forty-five trillion, six hundred seventy-eight billion, nine hundred eighty-seven million, six hundred fifty-four thousand, three hundred twenty-one."

"Okay, Matt," he said moving over to another patch of sand. He scratched a series of boxes. "How many total squares are there here?"

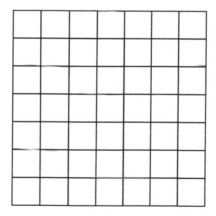

"Assuming those are actual squares?" I thought I was being funny, since, technically he couldn't draw perfect squares with exact ninety-degree angles and all of the sides perfectly equal.

He exhaled a puff of air, "Of course," he said. "Come on, Matt, this is important," he added, snapping his fingers eight times.

Math is about numbers, of course, but as I mentioned, it's also about patterns. You might even say that numbers are patterns, or patterns are numbers. Like the previous problem, some people might just give up or try to answer, "A lot." But start small to find the pattern and then apply it to the whole. Notice that if you just count the smaller squares, there are forty-nine of them, because there's 7×7 of them. Then there's the large square that forms the perimeter of the entire set. After that, there really are "A lot." But again, there is a pattern operating here.

Since looking for patterns is usually easiest when you start small and work your way up, start with a single square. How many total squares are there? Well, one, of course. How about the next largest set of squares? Two squares makes a rectangle, three make an "L" shape (or a longer rectangle). Remember the square numbers?

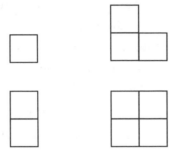

Four squares is the next time a set of squares can form a larger square because four is the next squared number 2^2 (2×2). In such a square there are four small squares and the larger one that forms the perimeter. That's a total of $4 + 1 = 5$.

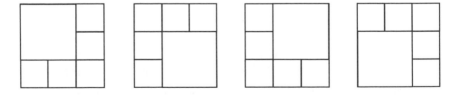

The next squared number is nine, since it's 3^2, or three times three. In this case there are nine smaller squares (1×1) each; four medium squares (2×2) each; and one perimeter (3×3). That's a total of $9 + 4 + 1 = 14$. Are you seeing a pattern yet? One more example should seal it.

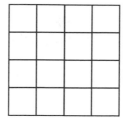

In this case, there are sixteen of the smallest squares (1×1) each; nine of the next smallest (2×2) each; four larger squares in the corners (3×3) each; and one perimeter (4×4). That's a total of $16 + 9 + 4 + 1 = 30$. So, the pattern is that we're adding squared numbers. The figure that Maglio drew had seven squares on a side.

So the pattern extended out would start with 7×7. The total would be:

$49 + 36 + 25 + 16 + 9 + 4 + 1 = 140$. Mentally rearrange that series of numbers: combine the 49 and the 1 to make 50, the 36 and the 4 to make 40. Then add the 9 and the 16 to make 25 and combine those with the original 25 to make an even 50. So it's really $50 + 40 + 50$.

"One hundred forty," I said.

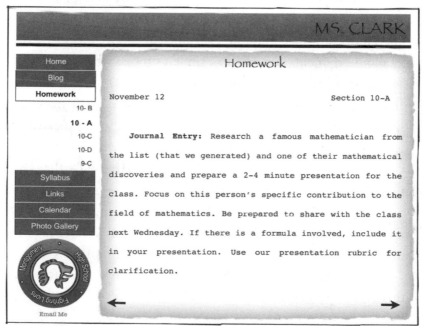

MS. CLARK

Home
Blog
Homework
10-B
10 - A
10-C
10-D
9-C
Syllabus
Links
Calendar
Photo Gallery

Email Me

Homework

November 12   Section 10-A

    Journal Entry: Research a famous mathematician from the list (that we generated) and one of their mathematical discoveries and prepare a 2–4 minute presentation for the class. Focus on this person's specific contribution to the field of mathematics. Be prepared to share with the class next Wednesday. If there is a formula involved, include it in your presentation. Use our presentation rubric for clarification.

"Okay, that was some pretty easy stuff, Matt. Let me make things a bit more interesting. What is the sum of the first two hundred whole numbers?"

There is a story that the famous mathematician Carl Friedrich Gauss was once asked in elementary school to add the sum of the first hundred numbers. Supposedly, the teacher gave the problem to the class to keep the students busy for a while, but Gauss saw a pattern and answered within mere moments. Like Maglio's last two questions, this calls for finding a pattern and deriving an equation. It's easy to get lost in numbers, so it almost always helps to start small and then build up toward finding a pattern.

Start with a small series, such as the whole numbers 1 to 10. Adding them in a series can be done as $1 + 2 + 3 + 4 + 5 + 6 + 7 + 8 + 9 + 10$, which isn't too bad to do mentally. But this summation would not work well for the sum of a series up to two hundred. Even with superior mental math, the process of adding every single number would take a long, long time. Math builds upon itself so that we can constantly take shortcuts to arrive at answers. For example, we first learn addition and might carry out the sum: $3 + 3 + 3 + 3 + 3$. When we then learn multiplication, we can shortcut this as 3×5. Later we learn about exponents and can do a similar process where $4 \times 4 \times 4 \times 4 \times 4$ can be written as 4^5 instead.

Back to the series $1 + 2 + 3 + 4 + 5 + 6 + 7 + 8 + 9 + 10$. Instead of adding them in their exact order, we can rearrange them to make them easier to add. For example, maybe you notice that $9 + 1 = 10$ and $8 + 2 = 10$. In this manner, we can purposefully select pairs that have rounded sums. These particular combinations are not what Gauss noticed, though. What he saw were that the end pairings created equal sums. In our 1 to 10 example, these would be $1 + 10, 2 + 9, 3 + 8, 4 + 7$, and $5 + 6$. Notice they all add up to 11, and since we were adding pairs of numbers, there are exactly half as many pairs as the total series. In this case, five pairs totaling eleven, or $5 \times 11 = 55$.

In the Gauss story, he noticed that the end pairings added to 101, such as: 1 + 100, 2 + 99, 3 + 98, 4 + 97 and so on. Again, because we are matching pairs, there are half as many as the total, in this case 50 pairs of 101. This is really 50 × 101, which can be done mentally as 50 × 100 plus 50 × 1, or 5050. It's only slightly different when there are an odd number of numbers, but I'll let you work that out. Remember, look for a pattern—and start small.

For Maglio's question the pattern holds true because there is a formula involved, or I should say there's a formula *because* there's a pattern involved. The pattern here was simply an end pair of 201 (200 + 1, or 199 + 2, etc.) times the number of pairings, or 201 × 100, or 20100. The formula in this case is n(n + 1) / 2, where n is equal to the number of numbers. Other variations on this formula exist for more complex patterns, such as the sum of a series of odd numbers or a sum of multiples of six.

"Twenty thousand one hundred."

Maglio spent most of the day testing me with a variety of problems. There were questions on finding the cube root of a number, some questions based in trigonometry, a few were about practical math applications, and he even threw in a few history questions. I had to answer questions on number theory, classic questions with trains approaching each other at varying rates, differential equations, and much more. It was by far the most grueling day of testing I'd ever had—way worse than the SAT or the ACT.

Every two hours or so he let me take a break so that Kelsie and I could walk down to a nearby stream and I could splash some water on my face. We filled up a garbage bag with water each time to replenish the supply at camp, keeping the water bottles for emergencies. Maglio always had tea ready for us in the coffee can when we returned to camp and used one of Kelsie's glass Mason jars to drink some down. When I got hungry, I gulped down a powerbar.

Late into the evening, as it started getting dark, Maglio stumped me on a surprising question.

"This has been pretty impressive, I have to say, Matt." I noticed he was walking in figure eights again. What was he doing, and what was with all the eights? "Perhaps a different approach is needed. You have all the standard mathematics in place—I will give you that." His eyes narrowed and he nodded his head a few times, "Hmmm, maybe … maybe this will work."

He grabbed our pine branch eraser and scratched out the definite integral problem I had just worked out in the dirt. He then pulled a mangled billfold out of his pants pocket. He reached in gingerly and took out four pieces of thick red cardboard. He bent down and arranged the pieces on the ground. There were two triangles and two trapezoids.

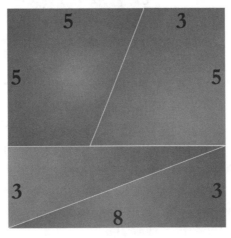

"Okay, Matt, these form a square that is eight centimeters on a side."

More eights? I looked over to Kelsie, but she just shrugged. I quickly went through the recent problems Maglio had given me in my mind. I didn't do a statistical analysis or anything, but there had been an anomalous number of eights in the answers.

Maglio continued, "The area is sixty-four square centimeters, correct?"

"Yeah ..."

He flipped the pieces over, "Now if I rearrange them into a rectangle like this, what is the area now?"

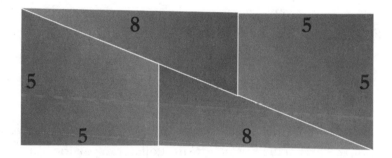

A sense of doom crept over me. I had a feeling I was in trouble considering he was carrying those cardboard polygons in his billfold. "Um, sixty-five square centimeters."

I mentally went through the formulas for the areas of triangles and trapezoids. Then, in a panic, I double-checked rectangles and even squares. Everything seemed to check out and yet something wasn't right. The pieces only got rearranged, so the area should not have changed.

And then the question I was dreading: "Can you explain this?"

My next thought was that this was a trick problem of some kind. Our teacher, Mr. Nguyen, was famous, or infamous, for giving us tricky thinking problems. He once gave us one called *The Motel Problem* that had many of us stumped for a long time.

The Motel Problem
Mr. Nguyen

Three men walk into a motel and ask for one room. The desk clerk tells them a room is $30, so each man pays $10. Later, the clerk realizes he made a mistake because the room should have been only $25. He calls the bellboy over and asks him to refund the $5 to the three men. The bellboy, not wanting to divide the $5 three ways, only refunds each man $1, keeping the other $2 for himself.

So, each man paid $9 towards the room and the bellboy got $2, totaling $29. But the original charge was $30, where did the extra $1 go?

The trick with the motel problem is that the so-called solution is misleading. There actually is no missing dollar if the math is done correctly. It is important in this problem to track what is really happening with the money, which can be made clear in two different ways. The desk received $30 and then paid out $5. From the $5, the bellboy kept $2 and the men received a refund of $3. So the math checks out. Another way to view this is that the men paid $25 for the room, the bellboy kept $2 and now they each have $1. Again, it checks out.

The confusion with this problem is that the two dollars that the bellboy kept was coming from the original money paid at the desk and should be subtracted from the $30. When the problem poses that the men paid $9 each, this is true, and they each have $1 that got refunded. Three times $10 equals $30. So, once again that math checks out. So, the desk has $27 - $2, or $25, which was the proper price. The men have the $3, which was refunded and the bellboy has the $2. So the math works out.

"Matt? Are you still with us?" Maglio said eagerly.

"Yeah, sorry."

"Well, can you explain this?"

I shook my head to try and clear my mind. Answering questions all day was tiring to say the least. The motel problem had me confused for days pondering how to account for the money. All along there was no missing dollar, and now I figured this new geometry puzzle was something similar.

I worked through everything again: the areas of the two triangles, the areas of the two trapezoids, the area of the final rectangle and the initial square. Everything seemed to check out…except for the fact that there was a mysterious extra square unit created by the rearrangement. What bothered me the most was that the area of the two triangles (12 square units each) and the trapezoids (20 square units each) should still add up to 64 square units. But I also couldn't deny that a rectangle with sides 13 units and 5 units has an area of 65 units squared. What was I missing?

I started getting suspicious that something was going on with the angles. Perhaps the rearrangements weren't perfect somehow, which would mean using calculus or trigonometry.

"Time is up, Matt!" Maglio snapped.

"I … but … I …" I muttered.

"You've been staring at the puzzle for the last eight minutes. Time is up."

I took a deep breath and closed my eyes. At least the tests were over. I felt like I did pretty well considering Maglio had been asking me questions most of the day.

Maglio picked up the four shapes and put them back in his bill-fold. He turned to me with a ridiculously big smile, "Good ole Lewis Carroll."

"The *Alice in Wonderland* guy?" Kelsie blurted.

"The same," Maglio stated.

"So, I just got stumped by a children's author?"

"I'm sure you will work it out," he replied. Maglio patted his pocket. "A lot of people do not know he was a mathematician as well. He published several mathematics books under his real name, Charles Dodgson."

"So," Kelsie continued, "he wrote novels and math books? That's just weird."

"So, what's the answer?" I inquired.

Maglio bowed obnoxiously, "Mathemagicians never, never, never, never, never, never, never, *never* reveal their secrets." With that, he turned and walked away.

19

Dreaming

Kelsie and I ambled back to the circle of huts and I collapsed onto the most comfortable log I could find. Between the lack of sleep and the grueling mental tests, I was practically drooling on myself.

Kelsie added a few more logs to the fire. "You look exhausted. Why don't you go to sleep?" she said poking the coals. "I'll stay up a while longer." She seemed finished, but then added, "I have a few things I want to do."

"And leave you with Crazy Eights?"

She shrugged.

"Where do you think he went, anyway?" I asked in between yawns.

"He headed north-northwest and ..."

There was a long silence where I felt my eyelids getting heavier and heavier. The next thing I knew Kelsie was sitting on the log next to me. "Matt, did you hear what I said?"

She poked me in the ribs.

"Wha ..." was about all could manage. I couldn't believe how tired I was.

"I was saying: I think I figured out why Fifty-Seven is hanging around near Maglio." She tilted her head. "He kept saying he loves ketchup and I kept wondering how that connects to numbers. Heinz 57 Ketchup. I'm betting that Fifty-Seven is Maglio's favorite number because he loves ketchup."

I think I nodded off again because the next thing I knew, Kelsie was shaking me. "Matt, what did you have to eat today?"

I shook myself and then mustered (sorry about the pun) the energy to stand up. "Is that supposed to be funny?"

"No. I'm just wondering."

"Just powerbars, some berries and some tea."

"Hmm. Okay." She was thinking deeply about something. "I think it's time for you to hit the hay—or the pine branches."

I nodded and meandered over to my hut and crawled through the entrance. I was so tired I just pulled the sleeping bag over me and fell asleep almost immediately.

I thought for sure I was going to have a nightmare about that geometric puzzle, but instead, I had a fascinating dream. I was standing alone in the middle of a barren field. In the distance a tiny dust cloud appeared and began moving toward me. As it got closer, I saw that it was a black eight-legged horse—perhaps Odin's horse, Sleipnir from Norse mythology. It sped toward me, galloping terrifyingly fast right at me. In the dream, I felt powerless to move and I just stood there in its path. The ground rumbled as its eight legs pounded the earth and kicked up dust. As it came closer, I could see that its eyes were dilated and it was breathing hard. I somehow sensed it was running away from something as much as it was running toward me.

Just as I felt like it was going to run over me, it came to an abrupt stop, the kind that would only be possible in cartoons and dreams. The horse leaned forward, panic in its eyes. I reached out to comfort it, but then suddenly it began to shrink until it became … a foot-tall figurine of a horse with eight legs rearing up on its hind legs. A chessboard about eight feet across materialized next to us and the Sleipnir figurine hopped onto the edge of the board. As it landed on a white perimeter square, the board morphed and became three-dimensional. Each ring of squares around the chessboard became the separate level to a small four-tiered tower, with the middle four squares at the peak.

The figurine then hopped up onto a black square in the next ring. As soon as it touched the black square, the figurine turned white. Then it hopped again, this time onto a white square and again the figurine changed to the opposite color: black. With one last hop, it landed on a black square and changed into a white figurine. It turned to me with a pleading look on its face. Bars rose up suddenly from the peak of the chessboard and formed a cage around it. Sleipnir tried to escape, but it couldn't jump high enough to clear the bars. It tried eight times, but every time it touched one of them, a blue spark flashed and a small black scorch mark appeared on its white surface. It looked at me for help again, but then suddenly I woke up.

The recurring number eight immediately caught me, especially because that was the number Maglio had been obsessing over. The horse had eight legs and it changed into a chess piece, and eight appears a lot in chess. There are eight squares on each side of a board, totaling eight times eight, or sixty-four. There are eight black pawns and eight white pawns. Then there are eight black power pieces and eight white ones. The king and queen can each move in eight directions.

The rings of the board that became three-dimensional also brought to mind the number eight. The outer ring of a chessboard contains 28 squares, followed by the next ring with 20, then 12 and finally the 4 middle squares. See the pattern? $28 - 20 = 8$. $20 - 12 = 8$ and $12 - 4 = 8$. Each ring has eight fewer squares. And, of course, they total 64, since $28 + 20 + 12 + 4 = 64$.

Finally, Sleipnir tried eight times to escape the cage, only to be scorched eight times.

There are eight heavens in Islam, the eightfold path in Buddhism, Christ spoke of eight beatitudes in his Sermon on the Mount, and there are eight symbols of the immortals in Taoism. There are eight parts of speech in English: adjectives, adverbs, conjunctions, interjections, nouns, prepositions, pronouns, and verbs.

Eight is a cubic number, meaning that it is perfect cube, in this case 2^3. It's considered by some to be the first real cube since one cubed is still equal to one and really hasn't changed. Eight as a cube appears in the *I Ching* where arrangements of broken or unbroken line segments create different symbols, and there are even eight times eight of them, or 64, in total.

Similarly, eight is the result of a series of doublings. One doubled is two. Two doubled is four. Four doubled is eight. If the process continued: 16, 32, 64, 128, 256, 512 and so on. This series produces one of the three types of even numbers. Known as "evenly-even" numbers, these numbers can be continuously divided into two parts until reaching the number two. "Evenly-odd" numbers can only be halved once. For example, 26 is "evenly-odd" because it can only be divided in half once, and the quotient is thirteen, which is an odd number. 14 is "evenly-odd" as well because it is composed of two sevens, and seven is an odd number. A number like 12, though, is "oddly-even" because it can be divided into two more than once, but not all the way down to the number two. This is because half of twelve is six and half of six is three, and three is an odd number.

What I'm getting at is that there are three types of even numbers based on what happens when they are halved: those that contain all even numbered halves (evenly-even), like 32 and 64; those that have a single odd-numbered half (evenly-odd), like 10 and 18; and those that contain both odd and even halves (oddly-even), like 24 and 48.

Types of Even Numbers

Example Number	Division by 2 Series	End Result	Type of Even Number
8	$8 \div 2 = 4 \div 2 =$	2	Evenly-even
64	$64 \div 2 = 32 \div 2 = 16 \div 2 = 8 \div 2 = 4 \div 2 =$	2	Evenly-even
26	$26 \div 2 =$	13	Evenly-odd
42	$42 \div 2 =$	21	Evenly-odd
12	$12 \div 2 = 6 \div 2 =$	3	Oddly-even
48	$48 \div 2 = 24 \div 2 = 12 \div 2 = 6 \div 2$	3	Oddly-even

This doubling or halving idea shows up in the human body as well. The most obvious example is in mitosis, when cells, strangely enough, multiply by dividing. What's really cool, though, is that twice around the base of your thumb is the approximate distance around your wrist. Double the wrist measurement and you have the distance around your neck. Then double the neck measurement and you have your approximate waist size.

Eight is made up of two circles, one on top of the other. One circle is usually written or drawn clockwise, the other counter-clockwise. This means that eight often has two different natures that are often seen in opposition to one another. Many people have noticed that an eight on its side is an infinity symbol and there are opposing natures to infinity: positive infinity and negative infinity (not to mention countable and uncountable infinity). Because of these two loops, eight is often a number associated with luck or fortune, either good or bad. Take the eight ball in playing pool, for example. Sinking it early will lose you the game, whereas sinking it after your seven solids or seven stripes will win the game. I wondered then about my dream and what it could mean for our situation.

Was it a sign of good fortune or bad luck?

20

The End of the World

That morning I crawled out of my hut feeling like I hadn't slept. My head was cloudy and my arms ached. My back was stiff from sleeping on the ground and I had a couple of sharp pains where various branches had stabbed me. I also had a kink in my neck from using one of Kelsie's windbreakers as a makeshift pillow. My appreciation for mattresses and fluffy pillows suddenly rose exponentially.

Kelsie was already up tending to the fire. She turned and smiled at me and I suddenly felt a lot better. She pulled two more powerbars out of her bag and held them out. "Breakfast of champions?" she said handing me one.

"Thanks," I stammered. I opened it and took a bite. It tasted like stale chocolate cardboard, but it was vastly better than acorn paste. We were both silent for a long time, until she finally said, "How'd you sleep?"

I laughed, "Like a log, or I should say, a branch." I stretched some of the cramps out of my body. "Thanks for the sleeping bag, though."

"No prob."

The dream was still vivid in my mind. I could still picture the figurine trying to jump out of the cage. "Did you have any, uh, weird dreams?"

"No," she shrugged, "but I didn't really sleep much either."

I kicked a pebble near my foot. "I had an interesting dream."

Kelsie laughed, "Was it about that puzzle?"

I shook my head, "No, thankfully, although I think I figured it out. No, the dream was about the number eight."

"Like, the number eight?"

I nodded again. "I think it ... I think it needs our help."

"What does that mean?"

"I don't know. I'm still puzzling all this out."

"*Puzzling* might not be the best choice of words right now, Matt."

"Eh, good point."

"So, tell me about the dream," she said.

I recounted everything to her as she grabbed a few logs from a large pile and added them to the fire. I noticed that the pile was new.

She noticed where I was looking. "Yeah," she tipped her head toward the pile of firewood. "I collected fire wood and fixed up a few of the huts that were starting to fall apart." She motioned to her backpack, "Dental floss to the rescue!" She took a bite of her powerbar.

"So, what do you think of this place, Matt? Maglio. Fifty-Seven. All the numbers. Look!" she said sweeping her hand in an arc. "The huts are six-sided and there are nine of them arranged exactly around this fire. And there's a circle of exactly three hundred sixty stones—I know, because I counted them this morning."

"Three hundred-sixty, huh? You have been up a while," I laughed.

"Why that number, do you think? I mean, it would take a while for someone to do that, but why?"

I shrugged, "It's the number of degrees in a circle."

"I know *that*. But why bother?"

I mentally went through what I could about 360. Perhaps the most important feature of three hundred sixty is that it is a *highly composite* number. This means that it has more factors than any smaller positive number. So, each highly composite number must have more factors than any previous composite number. Here's a chart with the first thirteen highly composite numbers. Notice each one has a greater number of factors. To make this clearer, 10 is a composite number because it has four factors: 1, 2, 5, and 10. But it's not *highly composite* because it doesn't have more than the smaller number six, which also has four factors: 1, 2, 3, and 6. Twelve, though, is a *highly composite* number because it has more factors than any previous positive integer.

Highly Composite Numbers

Number	Number of Divisors
1	1
2	2
4	3
6	4
12	6
24	8
36	9
48	10
60	12
120	16
180	18
240	20
360	24

The chart shows a bunch of numbers that turn up in a lot in the English measurement system, namely multiples of six, like 6, 12, and 24. Also notice that 24, 36, 48 and 60 (and many other highly composite numbers) are multiples of 12. As someone who loves the metric system for its simplicity and ease of multiplying or dividing by powers of ten, I was puzzled for years about the use of the number twelve. Why are so many things sold in dozens? Why are there twelve inches in a foot? Why are cans sold as six-packs, and 12 pence in a shilling? And what about a gross, which is twelve dozen, or 144?

Other familiar numbers are highly composite, such as 60 and 360. Sixty has 12 factors, while three hundred sixty has a surprising 24 factors and is divisible by all the numbers 1 to 10, with the exception of the number 7. This trait might be one reason why we divide a circle into 360 degrees. The Babylonians used a mathematics system based on the number 60, which may be why we have 60 seconds in a minute, 60 minutes in an hour and 360 degrees in a circle. There are many explanations as to why they used 60, but the fact that sixty is highly composite seems like reason enough to me. Also, three hundred sixty is close to the number days in a year, and many cultures used this number as an approximation.

An amazing highly composite number is 10,080—which is my second favorite number. It happens to be the number of minutes in a week (60 minutes per hour × 24 hours in a day × 7 days a week). It's an amazing number to test divisibility rules because it defies belief. It has an amazing 72 factors, making a number like 36, which has 9 factors, seem *hardly composite.* It's hard to believe, but 10,080 is divisible by 1, 2, 3, 4, 5, 6, 7, 8, 9, and 10. The smaller number 2520 has this property—and this is already cool—but 10,080 is also divisible by 12, 14, 15, 16, 18, 20, 21, 24, 28, 30, 32, 35, 36, 40—and still not done—42, 45, 48, 56, 60, 63, 70, 72, 80, 84, 90 and 96. You can work out the other 36 factors that pair with the numbers I just listed.

As highly composite numbers, 12 and 24 can be easily divided in numerous situations—unlike the number 10 from the metric system. To appreciate this difference we can compare the divisibility of 10 and 12. Imagine having a box of ten birthday cupcakes to distribute evenly out to friends without wasting any. Since the number 10 only has factors or divisors of 1, 2, 5, and 10, you would be limited in how they could be distributed. You could give two cupcakes to five people or five cupcakes to two people or one each to ten people. Besides keeping them all, those would be your only choices.

Add just two cupcakes to this box to make a dozen, though, and there are additional options, since 12 has factors of 1, 2, 3, 4, 6, and 12. This makes the number 12 significantly more divisible and, therefore, a better choice for things like marketing and selling stuff. The number of divisors also determines how things could be packaged for sale. If something came in packages of ten, then there would only be two main choices: one row of ten or two rows of five. With a dozen items, though, the packaging could be 1×12, 2×6, or 3×4, opening up an additional option.

Anyway, the main point is that although highly composite numbers like 12, 24, and 360 are a little harder to multiply than the metric system's number 10, they are easier to divide.

I shared all of this with Kelsie, including thoughts about the number nine and all of the associations to muses and personalities and stuff.

"So, what are you thinking?" she asked.

"I'm thinking: *Why are hot dogs typically sold in packages of ten, but buns are usually sold in eights?*"

"Matt!" she kicked dirt at me.

"You asked."

"Don't pin this on me."

I lowered my head in mock disappointment. "Sorry, I'm just hungry and achy and ..."

She held up her hand. "We're on an adventure, Matt! Quit being such a wimp!"

I took a deep breath, but didn't know what to say. I think maybe Kelsie felt like she had snapped at me a little too much because she began foraging through her backpack. She shifted the items around for a while without seeming to do much. I finished eating my card-board bar.

She looked up. "So, we're sitting in a highly composite circle with nine six-sided huts around a fire?"

"Pretty much," I nodded.

"But why?"

"Good question, Kelsie. I wish I had an answer."

"Give it a try," she said, playfully jabbing me in the ribs with her elbow.

I wasn't sure what to say, but some ideas began forming in my head, "Hmm … it's almost as if …"

"Hey!" Maglio's voice cut through the morning air, "hey, hey, hey, hey, hey, hey, hey! Good morning."

"Good morning," I replied. Kelsie just nodded.

"I'm sure you two have more questions." He sat down on one of the logs near the fire. He looked around, "I see someone's been re-decorating."

Kelsie was staring at Maglio's shoes for some reason.

"What was with all those tests?" I muttered.

Maglio seemed to ignore me. "Numbers are in everything. The air we breathe is a series of numbers put together in ordered ways. That tree over there is composed of molecules that are governed by numbers and those molecules are put together in numbered ways." He pointed to a weed growing by one of the huts, "The leaves on that plant over there, like many plants, grow in a spiraling pattern one hundred thirty-seven point five degrees apart." He held out his hand, "Our bodies are aggregated numbers governed by proportions and paradigms."

Kelsie looked back and forth between us a few times. "Are you two related?"

"Take the obvious polarity that defines the human body: we have venal and arterial blood that carry carbon dioxide and oxygen. We have a nervous system that relies on a potassium and sodium exchange. We have two ears, two eyes, two hands, two feet, two lungs and two hemispheres of the brain, and although we have one nose, we have two nostrils. We have thirty-two teeth, but that is two to the fifth power."

Maglio took a deep breath and opened his mouth to say something, but then stopped. He exhaled sharply. "We are walking collections of numbers, golems not of clay, but of numbers. So, understand that one of the eight or so things all this means is that people are connected to numbers in ways that they do not understand and cannot appreciate. Take favorite numbers, for example. Most people have a favorite number—and if you ask, by the way, it is usually four or seven. What people do not realize is that numbers have favorite people."

Kelsie turned to me, but I didn't know what to say, so I just shrugged my shoulders.

Maglio continued, "Here it gets complicated. Millions of people consider seven their favorite number. So, in a way, seven gets divided between all of them, making each individual connection fairly weak. Are you with me?"

We both nodded.

"But in your cases, not many people have 61 or 313 as favorite numbers, so you two are very strongly connected—Matt in particular."

Maglio closed his eyes and licked his lips. "Some numbers, though, do not have any supporters. They are not anyone's favorite and they cannot get people to like *them*. And that is where our problems begin."

"What do you mean?" I managed to ask.

Maglio took a deep breath, "Some of the numbers are, uh, dying." He turned and gave Kelsie a weird look, but I didn't think much about it at the time, since what he said suddenly struck me.

When I heard that numbers were dying, I first thought about Kelsie because her favorite number is connected to the idea of her father dying. I thought about Sixty-One dying and what that might mean to her. It would be crushing for sure. She might even jump to the conclusion that her father would die soon after without the concept of sixty-one there for support.

Kelsie piped up, "What do you mean *dying?*"

I looked over and she was biting her lip.

"I do not know what else to say," Maglio continued. "Some of the numbers seem to be getting sick or weak. Some of the numbers are missing and might already be dead."

"That might explain the flickering on Kelsie's phone." Maglio just frowned at me, so I continued. "When we first got here, her phone was glitching or something. All of the eights were flickering."

Maglio nodded and looked like he was about to say something when Kelsie interrupted. "*Dead*?" she pressed.

Maglio didn't respond for a while. Instead, he got up and walked around the fire. He pulled four pieces of firewood from Kelsie's pile and added them to the fire. "I have been wondering about these things for a while now. Here is the best analogy I have: numbers get their power because we believe in them, because we think about them. Like the gods in fantasy books, they are only as powerful as the number of worshippers they have."

Maglio returned to his seat. "So, take Fifty-Seven, for example."

I looked around, but didn't see his ghostly form anywhere.

"A lot of people think about ketchup and associate it with eating hamburgers, hot dogs, French fries and so on. So, Fifty-seven is a rather strong number in that sense. However, I am one of a select few people who consider 57 to be their favorite number, so it has a more powerful connection to me personally."

He smiled. "A number like one hundred eight doesn't mean much to most people. There are no consumer products based on that number. There are no sports teams named after it. There are no main units of measure to bring it up. People do not have 108 siblings, or 108 fingers and there are not only 108 channels on TV. So, there is nothing special about the number to make people think much about it. Even if someone had one hundred eight dollars in their wallet, the number itself would not seem that important. Are you with me?"

I nodded and realized that Kelsie was slouched over poking a stick into the dirt. "What about sixty-one?" she said quietly.

I tried to motion to Maglio to be careful with this subject, but he didn't see me.

"Well, sixty-one is your favorite number, right? That means that you and it have a special relationship. I imagine that as long as it has you, it will stay alive. Some of the other numbers are not so lucky."

Kelsie turned away. "But what about Sixty-One, is he okay?"

Maglio was about to say something, but then he turned to me. I must have given him a fierce look because he stopped altogether and started stuttering. "I, um, well, there is, I … Sixty-One is probably fine." Kelsie looked up from staring into the fire.

Maglio apparently wanted to change the subject because he turned to me. "Why is your favorite number three hundred thirteen?" He held up his hand. "Wait, don't tell me. A Pythagorean prime. An interesting choice, to say the least—if you can actually choose your own favorite number."

I tilted my head. "What do you mean?"

"Sometimes numbers choose you."

We were all quiet for a while watching the fire, but I wanted to return to the whole dying thing for Kelsie's sake. "Maglio, you said that some of the numbers are disappearing or dying. What do you think would happen?"

"I cannot be completely sure, but if my latest theory is correct, then the numbers that die here would cease to exist."

I gulped. There was that concept of nothing again.

Maglio continued, "Imagine, for example, if the number nine ceased to exist. Baseball would not be able to have nine innings or nine players on a team. We could not have nine o'clock or Ninth Avenue or anything to do with the number nine." Maglio turned his palms up and looked at his hands.

I nodded. Without nine as a digit there couldn't be a lot of other numbers, like nineteen or twenty-nine or ninety-nine.

Maglio continued, "Kids could not have a ninth birthday, calendars would not have any nines in them, and so on. If a digit like nine dies, the results would be catastrophic—to say the least!"

"What about numbers like," I paused to think about some fairly random numbers, four hundred thirty-three, or forty thousand three hundred twenty, or one million forty-three thousand six hundred and eleven?"

"Well, everything depends on whether they are truly significant to anyone or not. With the numbers you selected, perhaps nothing tremendously momentous would happen, at least at first. Those numbers are more uninteresting or dull and not many people think about them. I doubt, for example, if any of those are favorite numbers." He got up and paced around the fire. "Actually, forty thousand three hundred twenty is eight factorial, but not many people know that."

I was wondering how he knew that, but numbers work like that for me sometimes. They *speak* to me sometimes, and I don't mean literally like Fifty-Seven did. I remember once getting 8128 as an answer on one of our homework assignments for math class. As soon as I wrote the answer down, I thought, "Hey, that's a perfect number." I didn't know how or why it happened, but when I checked, sure enough, it is a perfect number. To this day I don't understand exactly what happened, but now that we were talking about weird things like alternate universes, numbers with personalities and numbers dying, I had a new revelation: Eight was trying to communicate with me— but for what reason?

Maglio continued. "If the numbers that make up an object are broken down or destroyed, then that object will crumble or cease to exist. The numbers that make up something usually just let things take a *normal* path. In other words, they don't usually interfere with what happens from day to day. But if they wanted to, they could, um, possess a part of that object and change it, alter it in ways that would utterly change or destroy it."

I jumped in. "So, with the four chambers of our heart, Four doesn't normally interfere with our lives, but if it wanted to, it could make our heart stop beating. So it could kill us, right?"

Maglio shook his head a bit, "It's not quite that simple. Four is made up of One and Three and twice Two. So they have a say in the matter as well. Four would have to overpower Three, Two and One. If it was strong enough, then yes, it could make your heart stop."

I could see where this was headed. That would make the number One the most powerful, followed by other simple numbers like Two and Three. One is special in that it is not evenly divisible by any number other than itself, so it wouldn't have to obey or overpower anything else. A prime number only has two divisors: one and itself, so those would be powerful in this sense as well. "That would make One the most powerful number, since it is *inside every* other number."

"Well, yes and no. It's not quite that simple either."

Kelsie exhaled out loud, "Is anything simple?"

Maglio smiled. "Every number comes from One. For example, one and one make Two. But Two is its own being. And although One and Two make Three, Three is its own being. They're separate entities, but they do interact with each other constantly."

"But that makes the first numbers stronger, right?"

"Well, yes, to be simple," he winked.

"Then all we have to do is to find One or Two and our problem's solved, right?"

"There's the problem, though."

"What do you mean?" Kelsie asked.

"No one has seen One, Two or Three in … a long time. And Four through Seven are missing, and Nine is supposedly sick or dying. Eight is the only single digit that I've seen around, but she's been, um, extremely shy. I can't seem to get her to help."

I thought about my dream with Sleipnir and the chessboard and all that and she didn't seem shy at all.

"You realize that we are fighting against time," Maglio said darkly.

"What do you mean?" Kelsie asked.

"We're in a battle to save the world. I thought you figured that part out already."

"Yes, but …"

Maglio cut her off. "With every passing moment, numbers like Nine are getting weaker and weaker. Without help they will die soon."

"What can we do?" I asked.

"For starters, we need more help." Maglio turned to Kelsie. "You said your favorite number is sixty-one, right?" She nodded. "And Matt, you said three hundred thirteen?" I nodded. "Well, you can start by finding them and bringing them back here before more numbers disappear or die and the world as we know it ends."

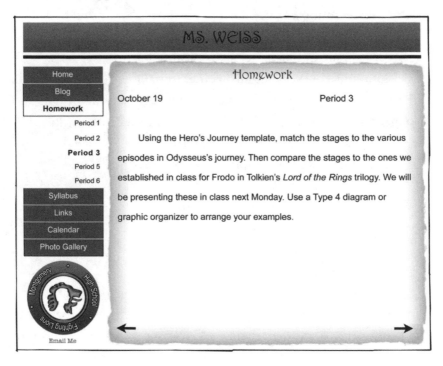

The realization that we had to save the world was troubling, for a number of reasons having to do with something called the Hero's Journey. In all of the fantasy and science fiction movies I've ever seen, the heroes are usually thrust against their will into some kind of grand adventure. In my case, check. Although, we did willingly go through all the codes, unlocked the journal screens and so on, we had no idea that we would get into this much … well, adventure.

And technically, Kelsie talked me into stepping through the teleportal.

Second, the heroes always meet up with some old wizard guy who gives them sage advice and guidance. Check. I guessed that would be Maglio, although the thought of comparing him to Gandalf or Obi Wan Kenobi was utterly depressing. Maglio was supposed to be our guide in this? I'd much rather have Yoda around.

Third, one of the other early stages in the hero's journey is the crossing of a threshold into the unknown or the different. Check. That was obvious here. The portal was an actual threshold and Kelsie I certainly did cross into the unknown.

After crossing the portal, there is a road of difficulty, a time of trial that tests the hero. Check. My grueling day of math with Maglio. In all the stories, the hero learns from the mistakes he or she makes during this period, and never dies at this point, otherwise the story wouldn't be very interesting. The trials after crossing the threshold are only supposed to test and train the hero, not kill him or her. But those are stories. Real life is different. What's to stop us from dying horribly painful deaths?

In the Hero's Journey, there is usually a sort of pre-quest to find a useful talisman or super item that could be used to vanquish whatever evil was threatening the world, something like a light saber or a magical sword. As yet, we had nothing.

Most worrisome, though, was the thought of what we were actually up against. In our situation we didn't seem to have a specific enemy. In every Hero's Journey there is an evil arch-nemesis like a sorcerer, a dragon, or at least some *thing* to fight. And in many of the stories that enemy often turns out to be a close relative. I laughed a little as I thought about engaging my mother in a supreme chess battle, or even better, my father in an epic television-watching duel. Then it dawned on me: I wondered if I would ever see them again.

Kelsie lifted her head, "So we're really talking about saving the world?"

"Well, yes," Maglio said, "to get right down to it."

21

Staying Busy

"Okay, here's what we need to do," Maglio said. "We need all the help we can get, and with your strong connection to Three Hundred Thirteen, I think we should start with him. Then use him to … er, then he can help us find other numbers."

"Like Sixty-One," Kelsie piped in.

"Yeah," Maglio muttered. He seemed distracted.

"Actually," I said, "I want to find Sixty-One first." Kelsie looked at me and smiled and I felt my face flush.

"What? What!" Maglio was clearly upset. "No, No! I want you to get Three Hundred Thirteen first and *then* we can get others to come. It will … actually, what did you say?" He turned and looked straight at me.

I leaned in and raised my eyebrows, "I said: I think we should get Sixty-One first.'" Whether he took my hint or he just didn't want to argue or whatever, Maglio simply nodded.

He exhaled sharply, "Okay, Sixty-One it is," he huffed, "be ready to go by tomorrow afternoon."

"Tomorrow afternoon?" I sighed. This adventure was already starting to get tedious. But at least I was with Kelsie.

Maglio turned, "There are several things we must do."

"First. It's possible we will be here a while, so Kelsie, as long as you have the knowledge and skills, we need you to set up camp for the long term. Food, water, *et cetera*."

"Okay, no problem," Kelsie said.

"Second. I will need time to calibrate the teleportal on this side. It is a long process with very little room for error." I nodded. That's certainly understandable since we wouldn't want to teleport into a solid wall or the middle of the ocean.

"Let's get started!" Kelsie agreed. I was about to complain, but she grabbed my arm, "It's okay, Matt," she winked at me. "I have a job for you."

The special job Kelsie had for me was *firekeeper*. I had to collect wood and keep the fire going. I can't believe how much wood it takes to keep a good fire going non-stop. Every time I thought I had plenty of wood, I seemed to space out a bit and suddenly we needed more.

Kelsie tried to convince me that my role as *firekeeper* was important and sacred, since keeping a fire going is much easier than restarting constantly. I could sort of understand that, but still, I was having trouble appreciating my position in the camp. While Maglio was out in search of additional numbers and calibrating inter-dimensional machines, Kelsie was busy hunting, gathering food and fresh water. I felt pretty useless sitting around the fire not really doing much.

I did get to use one of Maglio's nano-shirts to recharge Kelsie's iPhone. The cool thing was that I didn't even need the charger cord or anything. All I had to do was put the phone in the breast pocket and walk around for a while and somehow the shirt charged the battery. I was afraid at first to wear Maglio's shirt, but I had forgotten it was self-cleaning. When I put the shirt on, I expected it to smell like month-old socks, but it didn't smell at all. I also expected the nano-fibers to feel weird, but the shirt was like soft cotton. I can't wait until these become standard clothing styles!

Kelsie showed me how to boil water in a tree stump by using rocks. We heated the rocks up in the red-hot coals and then dropped them into the water. I didn't believe the water would ever boil because nothing much happened after four rocks, but when she dropped in the fifth, the water suddenly leapt to life and boiled.

There is something magical about watching water boil. I thought about *earth* (the stump), *air* (the air bubbles from boiling), *water* (the actual water) and *fire* (the heat in the rocks), but there was something truly amazing about watching the water go from a standstill to a complete boil so quickly. The magical moment it boiled made me think of something called *quintessence*, or ether, the "magical" fifth element that binds the other four.

In Chinese tea brewing and cuisine there are five stages of boiling. The first is called *shrimp eyes*, when the first tiny bubbles appear around 70° to 80° C. The next is called *crab eyes* when somewhere around 80° C streams of bubbles appear. At the *fish eyes* stage, around 80° to 90° C, larger bubbles form. At 90° to 95° C, the large bubbles form streams in what is called the *rope of pearls* stage. Finally, in the *raging torrent* stage, the water is at a complete boil. Considering that the last rock Kelsie dropped in caused the water to immediately boil, I could only assume that the average temperature must have jumped from below *shrimp eyes* all the way to *raging torrent*.

For the next sixteen hours, or so, Kelsie and I stayed around the camp to fix it up, while Maglio and Fifty-Seven came and left without much explanation. All Maglio said was that he needed to make preparations, which I took to mean that he was getting the portal ready, calibrating some machine, or something.

Kelsie was non-stop, bouncing from one project to another. Calling her Tigger seemed more appropriate than ever. At first it drove me crazy, but then I realized later that maybe this was her way to deal with the stress of our situation. Maybe by staying so busy she didn't really have to face what was happening.

Of all the things Kelsie made or did, by far the most intriguing was the figure four trap. She spent some time carving three sticks and then put them together to form the shape of a 4. The notches that she carved allowed the sticks to stay together in a delicate balance. Food or bait is put on the horizontal stick and the way the trap works is that the figure 4 holds up a weight of some kind. When an animal comes by and jiggles the bait, the figure 4 collapses, the weight drops and, well, dinner is served.

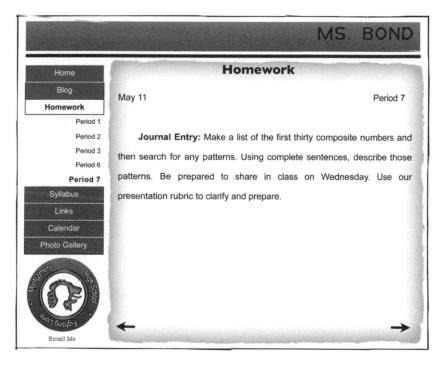

It's no surprise that the number four caught my attention. Four is an interesting number (although maybe all numbers are because of the interesting number paradox). One of the cool things about four is that it is the first composite number. As a figurate number, or figured number, four stones can be arranged to form the quadrilaterals, especially a square. But a new concept can emerge when four points are arranged as 3 + 1, where the fourth point is above or below the other three. What I mean is this: arrange the first three points into a triangle. Then add a fourth point above that triangle and imagine connecting them. What you get is a new concept: *volume*. In this sense, four is the first number that can create space to hold things.

Fire is an example of four being 3 + 1. Basically, fires consist of heat, fuel, oxygen and the chain reaction that occurs between the three. These four components constitute what is sometimes called the fire tetrahedron, which I think is cool (hot?) since a tetrahedron is sort of like the basic shape of a campfire.

There are, of course, the four operations: addition, subtraction, multiplication and division. There are four years of high school and four years of college. There are four main subjects: math (I put it first, of course), science, social studies and English. There are four passing letter grades: A, B, C, and D. School subjects often break down into four specific areas as well. Math, for example, breaks down into algebra, geometry, trigonometry and calculus. In science, there is earth science, biology, chemistry and physics. There are also four main degrees to work toward after high school: Associate's, Bachelor's, Master's and Doctorate's.

In music, you have to start with the Fab Four, otherwise known as the Beatles. There's also the Four Tops. There are four types of triads, or chords: major, minor, diminished, and augmented. There are the notes of something called the Tetrachord, or the first four notes of the Western musical scale: Do, Re, Mi and Fa. And there are the four main singing voices: soprano, alto, tenor, and bass.

There are four dimensions and four fundamental forces: electromagnetism, gravitation, nuclear strong force and nuclear weak force. Carbon and silicon have a four valency and these make up a huge amount of things on Earth. In the case of carbon, it helps to make up a lot of living things. Silicon makes up a lot of the Earth's crust and is found in materials like sand.

Here are more examples from my numbers collection:

1. Four bases make DNA and RNA.
2. Four chambers of the heart.
3. Four blood groups: O, B, A and AB.
4. Four main brain waves: alpha, beta, delta, theta.

5. Quadrupeds.
6. Four hits in baseball: singles, doubles, triples, home runs.
7. Four Horsemen of the Apocalypse.
8. The Four Gospels.
9. The Four Noble Truths in Buddhism.
10. The Four Rightly-Guided Caliphs.
11. Four elements: earth, air, fire and water.
12. Four directions: north, south east and west.
13. Four seasons: spring, summer, fall and winter.
14. Four suits of playing cards: hearts, clubs, diamonds and spades.
15. Four states of matter: solid, liquid, gas and plasma.
16. Four parts of structure: points, lines, area and volume.
17. Expressions: "Square deal", "Square meal", "Fair and square", "Back to square one", "Planted squarely on the ground."
18. Four corners of the world.
19. Origami always starts with a square.
20. The Tetragrammaton.

There are the Four Freedoms outlined in President Franklin Roosevelt's speech from January 6, 1941. These are: the Freedom of Speech & Expression, the Freedom of Religion, the Freedom from Want, and the Freedom from Fear.

And no list of fours is complete without four-letter words, but I'm not listing them here. I do wonder, though: Why four?

Considering that we were potentially on an alien world, I feel obligated to bring up a cool example of the number four. Many people have heard of the 1977 movie *Close Encounters of the Third Kind*, but they don't realize that there are actually four types of encounters. A Close Encounter of the First Kind is sighting aliens or their crafts. A Close Encounter of the Second Kind is actually finding specific proof, such as an alien artifact. The Third Kind is actually meeting aliens, so you can guess what the movie is about. And a Close Encounter of the Fourth Kind is actually visiting an alien world.

In our situation, Kelsie and I saw the ghost Fifty-seven, which was a close encounter of the first kind. Then we actually stood face to, uh, face with it, which is an encounter of the third kind. And if we did indeed pass through the portal to an alien world, then we also experienced a close encounter of the fourth kind. All we needed was to bring an artifact back.

What's really cool about things that come in fours is that there are often three things that are similar and one that is different. In other words, four often appears as 3 + 1. Go through just about any example of something that comes in fours and you can see what I mean. Take the elements, for example. There's earth, air, fire and water. Out of these, fire is the oddball. For starters, literally, in mythology fire was the element that had to be brought down from Olympus by Prometheus. In nature, the other three are prevalent in the environment, but fire must be sparked or created.

With the four directions, North is the unusual one, since compasses point that direction and North is usually used for orientation. In states of matter, plasma is the unusual one, since we normally only think about solids, liquids and gasses. In schooling, the senior year is the odd one out, since it is the graduation year. In sports, hockey is the only one played on ice. In the close encounters example, only the fourth kind requires leaving Earth.

Jamie and I made a game out of finding the oddball out of four things. We called it Quartering. We pick something like the four playing card suits and try to find some feature or trait that makes one of the suits stand out from the others. In order to do this you must have a context. For example, in the game *Hearts* you could obviously say hearts is the odd one out because the game is based on those for points, not to mention that the game is actually called *Hearts*. However, you could also say spades is the odd suit out because the queen of spades is worth 13 points. (Curiously, $13 = 1 + 3 = 4$).

What we quickly discovered was that some fours are easier to fnd than others. Take cars, for example. A lot of cars have four doors and you might wonder which one is the odd one out, but the driver's door is obviously the most used, so that's the different one. But take something like the tires. Which one is the odd one out? Jamie and I were stumped on that question for a long time. As it turns out (sorry about the pun) the wheels don't wear evenly, which is why tires are supposed to be rotated around the car every once in a while. Based on where weight is distributed in the car, one of the tires usually wears out faster than the other three.

We have played this game a lot in different places. Sometimes we play in study hall by looking at a group of four people sitting at a table. There's always something that three of them have in common and one does not. Sometimes it's that three people have glasses and one does not. Sometimes three are wearing the color blue and one is not. There's always something.

We started a similar game where we took two sets of fours and tried to match them together. For example, match the elements earth, air, fire, and water with football, baseball, basketball and hockey. Football could go with earth because the point of the game is to get territory and move down the field. Hockey might go with water, since the game is played on ice. Basketball can go with air since much of the game has to do with jumping and defying gravity, and as my dad pointed out, the famous Michael Jordan was nicknamed Air Jordan and His Airness. Baseball then goes with fire, and I'll leave that for you to work out.

This game or exercise is pretty cool because you have to think of reasons why things are connected together. According to one of our teachers, Dr. Peirce, this is *creative thinking* because we were making new connections between things. It's more fun than you might suspect.

231

Try, for example, to match the four directions with the four elements. Which direction goes with fire? Which with water? Air? Earth? The trick is that you have to come up with a reason why there's a match and that's half the fun. If you play enough times and match enough groups you can make really cool correspondence charts like this one:

	Sports Teams	Cardinal Directions	Terrestrial Planets	States of Matter	School Subjects
Earth	Football			Solid	Math
Air	Basketball			Gas	Social Studies
Fire	Baseball			Plasma	Science
Water	Hockey			Liquid	English

Anyway, I was fascinated by watching Kelsie work to make the figure four trap. She first searched for three good sticks. When she first told me about what she was doing, I assumed that there would be four sticks involved, but there are only three for the actual trap. But there's also the bait itself, meaning this was another example of the 3 + 1.

She carved special notches in the three sticks so that they would lock together. Once she had the pieces, she ventured out to find a place to set it up. I'm not sure what she was looking for exactly, but she just said, "A good spot." When she had done that, she came back to the fire area and wafted smoke over herself, the trap sticks and a log she had brought back.

"What are you doing now? I asked.

"I'm de-scenting myself before I actually set the trap up."

"I didn't know you smoked."

"Matt!"

"Sorry, just joking around."

I'm not sure if it was a coincidence, or if it was a feature of the world we were in, but she caught four rabbits over the next eight hours (4 × 2) … with a figure four trap. And rabbit, by the way, tastes better than chicken.

Later that night Kelsie and I stretched out on the grass and watched the stars for a while before crawling into our huts to go to sleep. Without light pollution, the sky was absolutely packed with so many stars that I couldn't even recognize any constellations. It was so magnificent that neither of us said anything for quite some time. After a while, I wondered what time it was. Then I wondered what day it was.

Then I wondered what month it was.

And what season it was for that matter.

Then, remembering where we were, I realized that I technically didn't even know for sure what year it was.

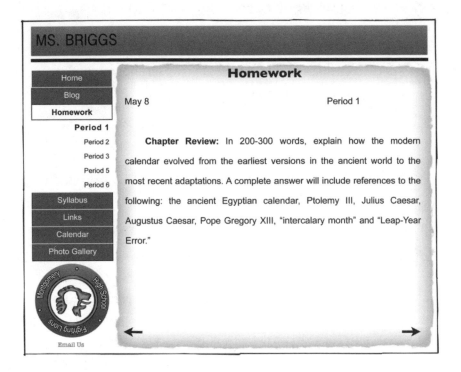

MS. BRIGGS

Home	
Blog	
Homework	
Period 1	
Period 2	
Period 3	
Period 5	
Period 6	
Syllabus	
Links	
Calendar	
Photo Gallery	

Email Us

Homework

May 8   Period 1

Chapter Review: In 200-300 words, explain how the modern calendar evolved from the earliest versions in the ancient world to the most recent adaptations. A complete answer will include references to the following: the ancient Egyptian calendar, Ptolemy III, Julius Caesar, Augustus Caesar, Pope Gregory XIII, "intercalary month" and "Leap-Year Error."

← →

It occurred to me that there might not be four seasons here in this world. The ancient Egyptians apparently thought of three seasons based around the flooding of the Nile. In modern times, of course, we mostly think of there being four seasons, but some indigenous cultures recognize five, six, sometimes even up to eight different seasons.

There was no specific reason that the months here would be the same either. For example, there are many types of calendars that track time in various ways. Some of them use the moon and its phases, some of them use solar calendars and mark the time it takes the Earth to go around the sun. Others like the Mayan calendar, are complex and use several variables. There are religious calendars that mark special holidays and events. Then we have things like school calendars, fiscal calendars and all sorts of other schedules that we use to govern time.

In our civil calendar, the Gregorian calendar, we divide the year into twelve months, totaling 365 days. We add another day to the calendar every fourth year, or "leap year," to account for the fact that our year is almost one-fourth of a day longer. I say almost because a year is more accurately 365.24, not 365.25 days. To account for this .01 difference, the years that are divisible by 100, but not by 400, are not leap years. For example, 1900 was not a leap year, but 2000 was.

You might not think a .01 difference is a big deal, and it's not for most purposes and to most people. But slight changes over time become very big changes. That .01 days becomes .1 in ten years; 1.0 day in 100 years, 10 days in one thousand years and so on. In two thousand years the calendar would shift 20 days. As I have mentioned, small changes or discrepancies can add up to big differences.

Although there's generally an accepted seven-day week now, with five days for work and two days for rest, there are many other divisions of days that have been used in history and throughout the world. The Soviet Union experimented with a five-day and a six-day week. The Etruscans had an eight-day week, while the Egyptians had a ten-day week. France experimented with a ten-day week around the year 1800 A.D. And the Mayan and Aztec calendars both had longer weeks of thirteen and twenty days.

All this was going through my mind as I looked up into the night sky. Wherever and whenever we were, at least I was happy being there with Kelsie. Plus, as she kept reminding me, we were on an adventure.

22

Finding 61

Kelsie was a mess thinking that Sixty-One was out there some-where and possibly in danger. I think she equated it with her father's life. As far as she was concerned, if Sixty-One died, then her father would follow. Given our entire discussion of what would happen if the numbers died, I certainly couldn't blame her. So both of us were glad when Maglio approached us the next day and said, "Okay, the first thing we need to do is find Sixty-One … for Kelsie's sake."

There was something in the way he said it that bothered me, but I was too relieved to think much more about it.

"Kelsie, you're going to be the key to finding Sixty-One because you have a stronger bond to him than either of us."

"Okay, but what do I need to do?"

"Well, essentially, you have to go hunting, er, *searching* is a better word."

"So, how do I do this exactly?"

Maglio was about to say something then stopped. He paused for a moment and his eyes flashed to me. I was suddenly scared of him and adrenaline coursed through me like an angry snake. Then his eyes softened and he looked thoughtful. "You just need to coax it out of hiding and convince it to join you. When I was searching for Fifty-Seven, I had a general feeling in which direction to travel. As a scientist, I had no explanation for this *feeling*, but it was there and it was tangible. Mothers have strong connections to their babies and entwined molecules have a resonance with one another, so I had these and other parallels for comparison." He stopped for a moment. "I simply had a general sense of where to go and I just kept following it." He trailed off.

We were all quiet for a moment and then Kelsie broke the awkwardness. "So … how long have you been here?" Kelsie asked.

"Well, that's hard to say. It all depends on how you want to track time." Maglio spread his arms and looked up at the sky, "Here it has been two hundred and three days since I first came through the portal. A day here is roughly the same as back home, although by my calculations, a year here is 408.32 days, rather than the 365.24 to which we are accustomed."

"How did you calculate that?" Kelsie asked.

"We brought some equipment with us in the first few …"

"*We?*" I inquired.

Maglio suddenly turned away. "Yeah, there were five of us originally."

"Where are they now?" Kelsie asked softly.

Maglio turned back to face us. "That's what I would like to know. When Fifty-Seven first told me that you had arrived, I thought it was some of my colleagues."

"What happened to them?"

Maglio exhaled deeply. "It's a mystery … and another problem to solve eventually."

We were all quiet for a while. "Maglio," I asked, "why haven't you gone back through the portal?"

He laughed, "Besides the fact that the world might end if we don't stop the numbers from dying? After Smolinski and Peters disappeared, I feared that something was wrong with the portal on the other side. Going back appeared to be a one-way trip. I decided to stay until I could save the numbers. I hoped that they would fix the portal on the other side and send help."

"And then basically we arrived?" I added.

"Exactly."

Kelsie looked confused. "I still don't understand what happened to them. I mean, what happened to the building, too? It was abandoned as long as I could remember."

"As to your school building," Maglio answered, "why it was abandoned is as mysterious as my missing colleagues."

We finished our conversation with Maglio and several things were clear to me. The first was that Maglio was right about saving the numbers. If we didn't do that first, then nothing else mattered—the world possibly ending and all that. The second was that he was also right about finding Sixty-One. Finding him was as good a place as any to start. Not only would Kelsie feel better about everything, but solving the mystery of where Sixty-One was would help us solve the greater mystery about all of the numbers.

I should have noticed it then, but Maglio had stopped doing and saying things with the number eight.

"So, are you ready?" I asked.

Kelsie laughed. "To find some numbers and save the world? Sure, no problem."

So we set out. Maglio had explained to us that we could find the numbers by listening to our feelings or instincts. I wasn't exactly sure what he meant, but I figured this was simply the next part of our journey.

"So, Matt, remember we were talking about the first three numbers and the anagrams?"

"Yeah, that was cool."

"What about some of the other numbers? I've heard you guys playing that game with the number four enough times. So, what about five?" she continued.

"What about it?"

"Well, what does it mean?"

Five is a fun number. When I think about five I think, "Five comes alive" because it's so vibrant and energetic. For example, there are five basic functions of biological life: respiration, reproduction, ingestion, digestion and excretion. People have five digits on each hand (4 fingers + 1 thumb), five toes on each foot and five main senses. We have five extremities: two arms, two legs and a head. We even have five holes in the face, which is pretty interesting. By the way, a lot of animals have four legs, a head and a tail, making them more like sixes.

In Jainism there are five classes of life based on the number of senses they seem to have. There are those that seem to have just the sense of touch, like vegetation. Then, there are those with touch and taste, like worms, followed by those with smell, touch and taste. More complex animals have sight, smell, touch and taste. Then higher animals, including humans, have all five senses.

In Shakespeare's day there were five wits, or inward senses, in addition to the regular five "outer" senses. There was common wit, or common sense, which is supposed to be the outcome of the other senses, or what the other senses had in common. The other four wits were (or are): imagination, fantasy, estimation, and memory.

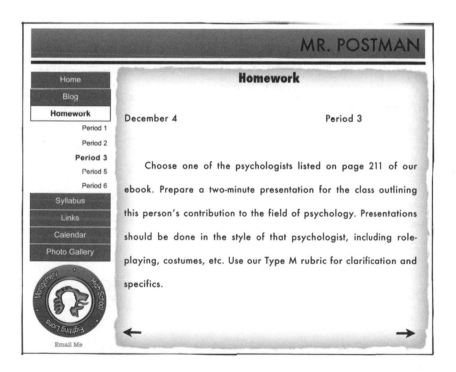

MR. POSTMAN

Home
Blog
Homework
Period 1
Period 2
Period 3
Period 5
Period 6
Syllabus
Links
Calendar
Photo Gallery
Email Me

Homework

December 4   Period 3

Choose one of the psychologists listed on page 211 of our ebook. Prepare a two-minute presentation for the class outlining this person's contribution to the field of psychology. Presentations should be done in the style of that psychologist, including role-playing, costumes, etc. Use our Type M rubric for clarification and specifics.

←   →

According to psychologist Gerald Heard, humans revolve around the number five. In his system there are five ages of man: infancy, childhood, adolescence, first maturity and second maturity. These go with his five stages of individuality: coconscious, heroic, ascetic, humanic, leptoid. Among other fives, he has five initiations as well: rebirth, catharsis, inspiration, illumination, and transformation. There are five basic levels in Maslow's Hierarchy of needs as well: biological, safety, love & affection, esteem, and self-actualization.

According to Dr. Peirce, there are five basic philosophical questions:

1. Cosmology: Who or what is in charge?
2. Ontology: Who am I?
3. Epistemology: How do I know what I know?
4. Soteriology: What is the purpose of life?
5. Eschatology: What happens after death?

"Well," I finally responded, "what are some things that come in fives, besides fingers and toes and stuff."

She looked around. "Um, white pine trees have needles in clusters of five—at least in our world. Your turn."

"There are the five pillars of Islam and Muslims pray five times per day."

"There are five positions in ballet."

"Really?" I said, "I need to add that to my iPad when I get back." I thought for a moment. "There are five elements if you include ether or quintessence."

"There are the five Books of Moses."

"There are five Great Lakes," I added.

Kelsie thought for a moment. "Aren't there five stages of mitosis?"

"According to our textbook, yeah. But I've seen different numbers for that."

"How can they be different? There are either five stages or there aren't."

"You'd think, but scientists can't agree on the exact stages I guess."
Kelsie shrugged her shoulders. "Hmmm ... there are five vowels."
"And five Platonic solids."
"Of course, you'd say that one," she said.
"There are five Olympic rings," I continued.
Kelsie jumped, "Ooh, that's a good one!"
"Why?"
"Well, it seems like a good example. The five rings represent the continents America, Europe, Asia, Africa and Oceania, and the five colors, or six with the white background, are to represent at least one color from every flag in the world."
"How did you know that?"
"My cousin was in the Olympics in China," she said.
"What event?"
"The four by one hundred relay, I think. His team didn't end up qualifying, but he said he had a great time there." I nodded, although the idea of being in a city packed with people and guards did not seem like fun to me.

We continued walking and listing examples of fives, like the five boroughs of New York, the Jackson Five, Euclid's Five Postulates, and many more. Every once in a while Kelsie stopped and steered us in a slightly different direction toward what I assumed was where Sixty-One was hiding or living or something.

Kelsie finally hit on a big one by bringing up stars. I'm not referring to stars in the universe—although interestingly, according to one of the first classification systems, by Angelo Secchi, there were five classes of stars. What Kelsie meant was five-pointed stars that are on the American flag, or the red star on Russian tanks, or the "five star" General of the Army. These stars can be seen all over the place for all kinds of things. What interests me, though, is that they have an unusual property in that the lengths of different sections of the stars are in something called the Golden Ratio.

The Golden Ratio or the Golden Mean is similar to pi in some ways, including the fact that it has a Greek letter name, phi. As I mentioned before, pi is the relationship between a circle's diameter and its circumference, between a curve and a straight. Phi, though, is a relationship between a whole and its parts. The way it works is this. Take a line segment and imagine splitting it up into two unequal parts. The longer part of that line segment we'll call "A" and the shorter side "B." The Golden Ratio is the place on that line segment where A + B divided by A is in the same ratio as length A divided by length B.

In other words: the Golden Mean is the ratio where the whole (A + B) compared to the larger part (A) is the same ratio as the larger part (A) compared to the smaller part (B). The amazing thing is that this relationship is always the ratio 1.61803, just as pi is 3.14159 (both rounded).

There are so many cool things about phi that I hardly know where to start. Remember the Fibonacci Sequence? If you take the numbers generated by the sequence and divide them by the previous number in the series, you will end up with the golden ration. So, take the numbers 0, 1, 1, 2, 3, 5, 8, 13, 21, 34, 55, 89 and so on. Divide each by the previous number: 0/1, 1/1, 2/1, 3/2, 5/3, 8/5, 13/8, 21/13, 34/21, 55/34, 89/55 and so on. At first the fractions are a bit erratic, but then they settle down and keep approaching 1.61803. It's amazing!

You can see phi in a lot of artwork. There's an idea that things that are in the phi ratio are more pleasing to the human eye. Of course, this is a bit controversial because of the cliché, "Beauty is in the eye of the beholder." The idea with phi is that people will generally find a shape more pleasing if it is in the Golden Ratio. So if you show a series of rectangles to people and ask which ones are nicer looking, then people will generally pick the ones in the Golden Ratio, also known as Golden Rectangles. For example, check these out. Which one is the best looking?

In my survey at school many people said that 'B' was the best looking, followed by 'D', with 'A' and 'C' way behind. These rectangles are set up in ratios of width to length. Notice that 'B' is in the Golden Ratio and 'D' is fairly close.

A 1 : 2   B 1 : 1.618   C 1 : 1.1   D 1 : 1.3

I know it's weird to think about nice looking rectangles, but it makes more sense when you think about those rectangles forming the walls of a cathedral or the façade of a temple.

You can make something called golden mean calipers by drawing out a pentagram and then connecting the various arms at the intersection points. These calipers are a lot of fun. You can measure just about anything and find the golden ratio in all kinds of things. A lot of plants and animals have the golden mean ratio in them. You can look at insects and their three body parts are often based on the golden mean. There are plenty of examples in the human body. Your finger bones or joints are in the golden ratio. The distance from your elbow to your wrist and from your wrist to your fingertips is in the phi ratio. Your navel divides your total height in the golden ratio, just as your eyebrows divide your face.

Spirals are related to the golden ratio as well. You can generate "golden spirals" by first creating golden rectangles, where the length of the longer side compared to the width of the shorter side are in the golden ratio. For example, if the shorter side is ten inches, then the longer side is 1.618 times that, or 16.18 inches. By creating a series of such rectangles and connecting the points with curved lines, you get a golden spiral.

By the way, there's also a Silver Ratio, which is produced by using something called Pell numbers.

Anyway, I was about to share some of this with Kelsie when she stopped suddenly. "Matt are you seeing this?"

I shrugged, "Umm, that depends on what you mean."

"How many birch trees do you see there?"

I followed her gaze and estimated. "Like fifty or sixty."

"How many *exactly*?"

I counted them. "Sixty-one. Hey! Sixty-one? What does that mean?"

"Birches are my favorite."

"Okay … but what does it mean?"

"I feel kind of weird. I think we're getting close." She took a few quick steps forward. "Hold on a sec. Do you see that pile of leaves?"

"Yeah."

"Watch." Kelsie stopped, closed her eyes and concentrated. The wind picked up and blew a single leaf onto the pile. "Now count them."

"Whoa! Sixty-one. Did you just control the wind or something?"

"No, it was more that I just *wanted* there to be sixty-one leaves there. I sort of willed it to happen."

"That's cool! It must be Sixty-One. We're close then."

"Kelsie, I have a *really* important question to ask." She stopped and looked at me. "Can you make sixty-one hot dogs appear?"

"Hot dogs? Seriously?"

"Yeah, I'm hungry."

"I know, but *hot dogs*?" I asked.

"What's wrong with hot dogs?"

"You mean, besides *everything*?"

"I know they're not exactly good for you, but I'm *so* hungry."

"You're always hungry!"

"I know, but right now I think I could try for a hot dog eating record."

"What's the record?" she asked.

"Something like sixty-eight in ten minutes."

"*Eyew*, that is so gross!"

"I know, but still …"

She placed her fingers on her temples and concentrated for a moment. "Nope, sorry, Matt."

"Oh well. Can I get another power bar at least?"

"I'm saving those for emergencies."

"Emergencies? This isn't an emergency? We're stuck in a different world looking for a number to help save our world!"

"I know, but there's more rabbit back at camp."

"Uhhh. I'm not going to win this, am I?"

"Nope," she stated. "Let's just find Sixty-One and get back."

We kept following Kelsie's instincts and came across several more collections of sixty-one things. We found a rock with sixty-one ants on it (they all had seven legs), a tree with sixty-one branches, a flock of sixty-one birds (which were hard to count), and a mound with sixty-one rocks. Out of curiosity, I picked up one of the rocks and it had sixty-one spots on it. After the mound, Kelsie took sixty-one steps and stopped in a small glade surrounded by sixty-one bushes.

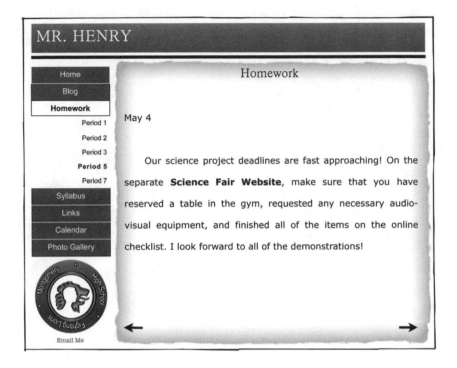

I started thinking that an aerial view of where we were might be really cool. I wondered how powerful Sixty-One actually was to affect the environment. I pictured being in an almost fractal landscape where anywhere and everywhere there were sixty-one things in ever-expanding connections. Something like sixty-one blades of grass in sixty-one clumps around the glade, surrounded by sixty-one bushes and sixty-one trees. On each blade of grass maybe sixty-one blemishes; each bush with sixty-one ants; and each tree with sixty-one birds or something. Zooming out: sixty-one glades (each complete with all the grass, bushes and trees), in sixty-one wooded areas in sixty-one valleys in sixty-one … well, you get the idea.

What I pictured wasn't exactly a fractal because it wasn't self-similar enough. For my eleventh grade science project I used some computer programs to generate different types of fractals. Fractals that are *exactly self-similar* are ones that appear the same no matter what scale they are observed. For example, using an iterated function, or a mathematical equation that uses a kind of feedback system, you can generate fractals that demonstrate the exact same patterns, even under a microscope. In other words, they keep repeating themselves in similar ways no matter if you zoom in or zoom out.

Fractals that aren't exact are *approximately self-similar*. This means that they appear quite similar, but aren't perfect representations of themselves. A lot of things in nature are approximate fractals and are even more similar than what I was picturing in our surroundings. Snowflakes, crystals and river networks appear self-similar. Again, this means that zooming in or zooming out, the parts look quite similar. Coastlines, mountain ranges, and lightning are practically exact fractals, or *fracticals*, as John dubbed them. Cauliflower and broccoli are two more examples found in nature.

So imagine walking up to a very small stream. Kneeling down in the mud you look and see a tiny spiraling pattern about a square inch in size or so. Then, standing up, you might see that the square inch section was part of a larger pattern made up of other swirls. Then get in a helicopter and look at the larger river system in the local area, and they might look just like the one square inch swirl. Then download a satellite image to view the entire river network, complete with tributaries, and it still might look the same as the original swirling pattern.

Anyway, Sixty-One wasn't apparently influential enough to create a wicked cool fractal landscape complete with 61 of everything repeating out from this glade because I noticed that the background was dotted with only a few oak trees, instead of sixty-one.

She looked around, "It's here somewhere."

"*He's*," I corrected.

"What?"

"*He's* here somewhere."

"Oh, yeah," she said. "*He's* here somewhere."

"Where though?"

"If I knew, I wouldn't be looking around."

"Good point." I scanned the bushes around us, but didn't see anything unusual. Well, if you don't count that we were in a different world surrounded by things in groups of sixty-one.

"There," Kelsie said pointing.

I turned to where she was pointing and saw Sixty-One float out from behind a birch. Like Fifty-Seven, he was made up of fluctuating black and white squares and rectangles. And he was vaguely human shaped, and reminded me of a bad ghost costume on Halloween. Sixty-One had violet eyes, though, and seemed more translucent or less dense or something. I got ready to snap some QR pictures, but a pattern never formed.

Kelsie concentrated a moment and he slowly came out. Now, I know I'm not exactly a judge of ghosts or anything, but it seemed scared or at least nervous.

Kelsie crept forward and then knelt down and held out her hands in a non-threatening posture. "Come on," she said, "we won't harm you."

"I don't think he can understand us yet."

"How do you know?" she retorted.

I was about to argue, but realized that technically she was probably right.

"Come on," she repeated.

Sixty-One shimmered in place and then slowly floated toward us. As it grew closer, Kelsie reached out her hand.

"Kelsie!" I called out.

"What?" she said, pulling her hand back.

"You don't know what could happen."

"Adventure, remember?" she said in a *Matt, you're being a wimp* voice. Before I could say anything else, she reached out and touched Sixty-One. Her hand passed inside and refracted a bit, like putting your hand in water.

She giggled and pulled her hand out. "That tickles!"

"What's it like?"

"It's like … it's fizzy or bubbly, like touching soda or something."

Kelsie stood up slowly, "Come here and try it, Matt."

I walked up and noticed for the first time that Maglio's "cloud of potential energy" description was not nearly accurate enough. I had not seen Fifty-Seven up this close, so I was surprised to see that Sixty-One was a lot more "alive" than even I had supposed. He shimmered and pulsated like a computer constantly refreshing, and he expanded and contracted like he was breathing. And, although his eyes didn't noticeably move, I had the distinct feeling that they were looking at me. I hesitantly reached out my hand, and sure enough, it felt fizzy like soda or carbonated water.

"Weird," was all I could say.

"I know! Isn't it cool!" Kelsie said, bouncing in place.

We stayed there for a bit while Kelsie tried her best to communicate with Sixty-One. We tried talking with him and waited for QR code patterns to emerge, but nothing seemed to work. Kelsie wasn't successful at teaching him much, but she did get him to follow us consistently.

"Now that you're a ghost whisperer, can I get those hot dogs? Or at least the buns or something?"

Kelsie laughed, "You'd eat sixty-one hot dog buns right now?"

"I'd eat sixty-one cans of pickled cauliflower if they were around."

"Eww, that's disgusting!"

I was exhausted when we got back to camp. I'm not sure what iteration of sixty-one we had just walked, but it felt like sixty-one leagues, (approximately 183 miles), or sixty-one miles, or at least sixty-one kilometers (37.82 miles). I know I'm exaggerating, but we walked for a long time. A more reasonable estimate, I suppose, was that we had walked somewhere between sixty-one furlongs (7.625 miles) and sixty-one li (8.91 miles).

Maglio stood up just as we crested the hill, "Ah, I see you found him." I could have sworn Sixty-One backed up a bit when Maglio spoke, but I wasn't sure because it was only something small, like maybe sixty-one millimeters. "Here, I reheated some tea," Maglio said, handing us two wooden bowls that Kelsie had made.

"Yeah," was all that Kelsie could muster, which was more than I felt like saying. She reached out and grabbed a bowl of tea as well.

I took the bowl and downed the tea. It was lukewarm, but I enjoyed it immensely. Kelsie had made it with some plants she had collected, mostly evergreen needles.

Maglio looked happy. Really happy. "Okay, Matt, it's time for Three Hundred Thirteen."

"I'm way too tired to go looking for him. Besides, he's probably three hundred thirteen miles away or something. I'm hungry, too."

"Matt, this is important. We're up against time, here. Let's just try something. If it doesn't work, you can try again tomorrow."

"What do you want me to do?"

"I want you to bring Three Hundred Thirteen here," he said pointing in front of him.

"Um, how do I do that exactly?"

"Just *will* him here."

"Just try it," Kelsie said, "if it doesn't work, we'll get some sleep and start tomorrow."

I nodded. I guess it was worth a try. I started out by just thinking about bringing Three Hundred Thirteen to me, much like willing a hot dog to float through the air from a grill to land into my awaiting mouth. Then I pictured another hot dog floating through the air and intercepted by a bun. On its way to my mouth it rained ketchup and a little relish, just enough to deliciously coat the top of the hot dog, and then it came spiraling into my mouth. Mmmmmm.

And then I realized I was thinking about hot dogs.

So I shook my head clear and focused on Three-Hundred Thirteen. I thought of it as a hypotenuse, as a product, as a sum, and in many other forms. I thought of 313 hot dogs, which I counted carefully, and then imagined Three Hundred Thirteen as one of the ghost-things. Then I simply willed him to appear next to me.

I didn't expect it to work, of course. I think part of me didn't even want it to work, since I really wanted to go to sleep. But suddenly with just a minor stirring of the air, Three Hundred Thirteen was hovering next to me. Like Fifty-Seven and Sixty-One, he appeared as a gray cloud of black and white squares and rectangles in vaguely human form.

"Hmmm, I'm not surprised," was all Maglio said.

With a closer look, Three Hundred Thirteen seemed very much like the others, a swirling cloud of potential energy. However, he had violet eyes and was a lot more solid or opaque. I waited for a QR communication moment, but a specific pattern never emerged.

A burst of adrenaline coursed through me and I thought the moment was going to be a lot more exciting—after all, we had just passed our first Hero's Journey tests. Surely, with the help of Sixty-One and Three Hundred Thirteen, we would be able to gather together more numbers and save both worlds!

I'm not sure if summoning Three Hundred Thirteen exhausted me even further, or if I was just that tired, or what. But the excitement faded quickly. I vaguely recall thinking about how cool it was and I have a fuzzy recollection of crawling into my hut and falling asleep. Summoning Three Hundred Thirteen didn't seem very important for some strange reason. And, unfortunately, I didn't really think much about what Maglio had just said.

23

The Hero's Journey, Declined

I woke up and found myself gagged and bound. My legs were tied together and my hands were tied behind my back. And to make matters worse, the gag was horribly bitter—from acorn water or something. My head hurt—a dull buzzing like a headache only worse. I could barely keep my eyes open because even the dim light in the shelter made my head hurt even more. Kelsie was on the other side of the shelter, gagged and bound just like me.

"I see you're up now," Maglio said. He was no longer disheveled and distraught looking, but now was clean-shaven with new clothes and his hair pulled back into a ponytail. I almost laughed when I saw that he had a white pocket protector. Almost. Because clipped to the pocket was a large folding knife.

I had equated Maglio to the wise sage or wizard who is supposed to help the hero along the Hero's Journey, so you can imagine my complete surprise when I realized he was the enemy. I was suddenly remarkably disappointed in the Hero's Journey.

Maglio motioned to Kelsie, "I don't think she'll be getting up any time soon. Sleeping poison." He smiled and it was half *I'm going to give you a million dollars* and half *I'm going to feed you to rabid zombie sharks.* He nudged Kelsie with his foot and when she didn't move, he continued, "Matt, in a moment I'm going to take your gag off and then we're going for a little walk. Then you're going to help me with something. After that … well, we can deal with that when it comes."

He looked back to Kelsie and then stepped closer. As he did, he pulled a small club out from behind his back. "Matt, if you try to cause any trouble I might need to silence you for a while." He looked like he might enjoy using me as a baseball. He leaned down and then ripped off my gag.

"You drugged her!"

He shrugged, "Well, yeah."

I'd like to say that a blast of adrenaline coursed through me and I ripped apart the rope that bound my wrists, but unfortunately, struggle as I might, I could barely move.

"Where's Sixty-One?" I said, thinking about Kelsie's father.

"He's serving a greater good. Wait, that's not true. He's serving my good, which is all that matters right now." He paused for a while. "This is your lucky day, Matt. You and I are about to make history."

"What do you mean?"

"Oh, you'll see. Now, I'm going to untie your legs. Don't get any ideas, Matt. You're not the heroic type." That stung. Not the heroic type? I was here, wasn't I? He knelt down on one knee and set the club down. I thought about trying to grab it, but with my hands tied, I didn't think I could do much. Maglio untied the rope around my ankles and then grabbed the club and stood up. "Come on, we have a bit of a hike." And with that he pushed open the flap and motioned me outside.

"What about us saving the world?" I whined.

"That is such a cliché: the world ending. Why is it that in most science fiction and fantasy stories the heroes are always supposed to save the world from imminent destruction?" I shrugged because I didn't know what else to say. Maglio continued, "This was never about the world ending, although you two bought into that nicely. No, this is about making me very rich." He motioned me outside again. "Now, come on."

I hesitated and looked back to Kelsie.

Maglio followed my gaze. "She'll be fine when the poison wears off. You two didn't notice it against the bitterness of the acorns." He laughed, "You know, I really couldn't have set things up any better than when she had the idea to boil those acorns."

"Is she going to be okay? Because if you hurt her …"

"As long as you cooperate, Matt. Everything depends on you and Three Hundred Thirteen now."

That caught my attention. What was he up to? How could I stop him and get out of this place? Well, I didn't have any great ideas as yet, but an inkling of something was coming to me.

Fifty-Seven and Three Hundred Thirteen were outside waiting for us. As soon as I appeared, Three Hundred Thirteen floated over and hovered close to me. I tried communicating with him, but he didn't respond. Sixty-One was nowhere to be seen.

"You have to train them, Matt."

"Huh?"

"You have to teach them so that you can communicate, Matt. Unfortunately, you won't have enough time to do that."

Fifty-Seven took up a position behind us as Maglio led me up into the hills that overlooked the camp. In a lot of movies I've seen the bad guys always give away their "secret" plans because they're overconfident and feel that they can't be stopped, so I decided to press him. "So, what are we going to do?"

"You, Matt—or actually, I should say Three Hundred Thirteen—is the final piece to something I have been working on for a while now. I have to admit I am surprised it is a bratty kid like you that will be making me a very, very rich man, but whatever."

"Rich? How?"

"You expect I am going to tell you everything? Well, I am not. Let us just say that you are going to make Three Hundred Thirteen do me a little favor and ..."

"*Make*?" Something about the way he said that word made me cringe.

"You will see," he said with a devilish smile. He walked toward the hut I had come to think of as Calliope (epic poetry) and *peacemaker*, but then he kept walking. Fifty-Seven was behind me, apparently escorting Three Hundred Thirteen and me like we were prisoners.

We walked for a while and with each step I felt the tension rise. My legs were shaking and I noticed I was breathing heavy. I glanced back at Fifty-Seven and his random patterns and shiftings had accelerated. Maglio didn't say anything to me, but I could hear him muttering to himself, laughing and making strange whimpering noises. It dawned on me that Maglio was probably planning on killing us once I did this "favor."

I finally got so nervous that I put my head down and started counting my steps, something that I hadn't done since my younger days when I only ate prime numbers of hot dogs, even numbers of hamburgers, or square numbers of potato chips. I counted one hundred twenty-one steps and then Maglio stopped.

I'd like to pause and say that one hundred twenty-one is a cool number. It is a palindrome square number, but it's also a star number. Like other figurate numbers, star numbers can be formed into regular patterns, in this case centered hexagrams, or six-sided stars. Think of a Chinese checkers board, which actually has one hundred twenty-one holes in its six-sided star pattern.

One hundred twenty-one is also the sum of three consecutive prime numbers: 37 + 41 + 43. It's also cool because it is the only square number that is in the form of $1 + p + p^2 + p^3 + p^4$, where p is equal to a prime number. In this case, the sum of: $3^0 + 3^1 + 3^2 + 3^3 + 3^4$, which is $1 + 3 + 9 + 27 + 81 = 121$. One hundred twenty-one is also the number of points needed to win a game of cribbage. In the Dewey Decimal System, 121 is for Epistemology, which is something like studying how knowledge is created and where it comes from.

When I finally looked up, I found that I was standing in front of a small cabin, approximately thirty by forty-five feet. It was made of rough timbers and there was a single window next to a set of double doors. The pitched roof was made of wooden shingles, and jutting out from the top was some sort of weather station, complete with one of those spinning wind velocity things. Four equally spaced solar panels were visible on this side of the roof.

"You had a cabin this whole time!"

Maglio ignored me. "Get in there, Matt," he snarled. "Now!"

When I opened the double doors to the cabin, I expected to see a basic middle-of-nowhere cabin or hunting lodge: bunk beds, wood-stove, and rough-cut tables and chairs. Maybe there would even be an animal head or two mounted on the wall, a bearskin rug spread across the floor, and in this case, some scientific equipment.

However, the doors opened into the dining area of the cabin. Straight ahead was a small kitchenette and pantry. Fifty-Seven slipped in ahead of me and took up a position near a pile of messy crates with cans of food spilling out all around. My mouth watered to a disgusting degree as I stared at cans of baked beans, corn, carrots, vegetable soup and tuna fish, packages of spaghetti, cereal, granola bars and dried fruit. I had to wipe drool away from the corners of my mouth, even though most of the cans still had food in them and there were flies buzzing everywhere. Then the smell hit me—a cross between rotting vegetables and gym lockers—and I nearly puked. I thought my room was a mess, but this was disgusting.

I took a few steps forward, a sense of dread creeping over me. Three Hundred Thirteen floated in behind me. To the right was a rough-cut table and chairs. Playing cards were arranged at one end of the table in what was probably a Solitaire game. There was a small room to the left with the door open. I saw that there actually were bunk beds inside. A closed door in the back wall lead to what I estimated was the other half of the cabin.

I tried again to mentally communicate with Three Hundred Thirteen, but there was no response. At one point I thought that maybe a QR pattern had formed, but it could have been my imagination. I must have stopped to gawk at the food because Maglio shoved me from behind. "Keep moving, Matt!"

"You had real food too!"

"Of course! You think we would have been stupid enough to come through the teleportal unprepared? This is where we …"

"*We?*"

"Yes. *Superiors* gives the wrong impression and *colleagues* implies that they were not complete idiots. *Hindrances, nuisances* …" Maglio paused and looked around as if he were searching for a word. "… *Leeches* might be best." He gritted his teeth and looked at me in a way that would make a serial killer cringe. "They wanted to suck me dry of every innovative genius cell in my body. I have more intelligence in my left … ah, Maglio digresses."

He shoved me backwards toward the door in the back. "They are buried out back, you know—or more precisely, I just threw their …" He stopped and stared at me for an uncomfortably long time. "You do not get any of this, do you?"

I tried to say something, but all I managed was, "I … I …"

He shook his head. "This will make more sense to your feeble mind in a moment." He motioned toward the door. "I spent years," he grumbled, "on all of these inventions and for what?"

I had so many questions racing through my mind that I thought I was going to explode and send question marks flying everywhere. "But … but what about the numbers dying and saving the world, and … and … why'd you have Kelsie and I do all that work around camp?"

He shrugged, "Curiosity mainly."

"Curiosity!"

He shrugged, "It was informative to watch, I have to admit. After all, there is the slight possibility that I will be here longer than I wish." Maglio grinned, "But more importantly, I had to capture Eight—and thanks to you, I accomplished that. Now get moving!" He shoved me so hard that I lost my balance and nearly crashed into the pile of crates.

I suddenly realized that Maglio's whole appearance, starting with when we first saw him, his dilapidated look, the story of the numbers dying, the world coming to an end, everything he did was designed to catch us off guard and to make us sympathize with him. I had never suspected him as the enemy. He motioned me toward a door in the back. I passed a crate of pickled herring and wondered how it would taste.

24

Surprises

I opened the door to reveal the entire back room of the cabin, roughly 975 square feet, lit by five windows. The entire room was packed with what I first thought was just scientific equipment, like the lab back at school. But all the walls were lined with metallic cages, maybe forty or so total. Wires snaked out from each of the cages and led to a large metallic block in the center of the room. It looked like something called a Menger Sponge at about the third iteration and emitted a dull humming noise, like a generator. From the metallic cube a single thick cable snaked its way over the floor to another teleportal in the corner closest to me.

I scanned the room. All of the cages except for one had a number-ghost in them. They had various eye colors, and I hardly would not have been able to tell them apart except that the cages were labeled. Thirty-Nine and Twenty-One had yellow eyes, just like Fifty-Seven. Seventy-Nine had violet eyes, just like Sixty-One and Three Hundred Thirteen. Others were red: Nineteen and Ninety-One; and orange: Eleven. There were a few green ones, like one close to me, which was One Thousand One Hundred Eleven. Blue was represented by Fourteen and Seventy-Seven; and indigo by Thirty-Three. At the far end of the room I even spotted some with gold eyes, Eighty-One, and two with silver eyes: Eight and Six Hundred Two.

I saw Sixty-One in a small cage fluttering around frantically throwing himself against the bars. Each time he touched the cage, a blue spark flashed and he flinched back, only to try again. Eight was in the largest cage with the thickest bars and she was trying unsuccessfully to get out as well. The other numbers were strangely unmoving. There was an empty cage nearby labeled 313.

"I see you are beginning to work things out now," Maglio said. "Yes, I would have eventually captured one of the Ennead Numbers without your mathematical connections, but you certainly made things a whole lot easier. And getting Sixty-One and Three Hundred Thirteen? What a delicious bonus!" He smacked his lips.

My mind was racing, "Why Eight?"

"She was always the best bet. One, Two and Three are way too powerful. Four, Five and Six have too many connections. Seven is associated too strongly to good luck and heavens and is protected by … certain things. And Nine is too tricky and elusive, but that is another story. Eight was always the most probable."

Suddenly a few things made a bizarre kind of sense to me. Maglio hadn't been living in the huts at all. The three hundred sixty stones and the nine huts was some sort of lure or trap, perhaps because 360 is a highly composite number divisible by every digit except seven. The nine huts were for each one of the single digits. It dawned on me that Maglio had somehow been luring numbers to the camp and trapping them for months. I thought of my dream about Eight, and somehow we had been the tipping point.

"Matt, I'm going to use Three Hundred Thirteen as the last piece of a grand puzzle. A series of events is about to unfold that will make me the richest, most important, most powerful man in the world." He swept his arm in a wide arc. "Welcome to my investment portfolio! Each of those groups over there is calibrated to a specific lottery around the world. In just a short time, they will fabricate the winning numbers. That group there," he pointed to a group with Sixty-One, "is going to win the Super Powerball." He pointed to another group, "That one is going to win the largest European lottery. Those over there are going to modify the price of gold and silver, and those are going to affect the NASDAQ, and, well, you get the idea."

Maglio pointed to the empty cage, "Now, make Three Hundred Thirteen get in there!"

In all of the movies I've seen, when the bad guy has the upper hand, the hero always stalls, waiting for an opportunity or for reinforcements to come. "You know, you're not going to get away with this! You think people won't think it strange that one person just happened to win a bunch of lotteries at the same time? The odds are … the odds are …" I tried to think of something amazing to say. But all that came out was, "the odds would be bad."

"Bad? Bad?" Maglio laughed. "You have a way with words, Matt. Yeah, the odds would be astronomical if the same person won all of them—practically impossible and statistically improbable."

"They're going to catch you."

Maglio smiled, then motioned and, in an instant, Fifty-Seven flew over and in to him. Maglio's features changed instantly. His face thinned and his cheeks hollowed. A few clumps of his hair fell to the ground. The hair that remained on the sides turned gray and then white. People change over time, but they're usually still recognizable even many years later. Maglio had only aged twenty or so years, but something about the rapidity of the transformation made him appear completely different. If I had not seen the change take place right before my eyes, I never would have believed he was the same person.

"You see, Matt," his voice was hoarse and soft, completely unrecognizable. "Everyone will be looking for a different, younger *Maglio*, the one that disappeared a few years ago. They will have no reason to look for a fifty-seven year old—I call this persona Tony." He swept his hand around indicating all of the cages. Then he pointed to two other cages labeled 50 and 32, "I like Nan and Ruth as well, although I am still getting used to changing into women." Maglio turned and glanced to another cage labeled 19. "My favorite persona right now is nineteen year-old Robert. There is something wonderful about the energy and power of youth."

He turned back. "Anyway, Maglio digresses." Fifty-Seven popped out of him and his hair shrank back and darkened almost instantly. His cheeks filled out and in just seconds he appeared normal again, although I realized that I didn't know what normal was for him. He stretched his arms and clutched his fingers. "The amazing thing is that our fingerprints change, as though each of us is really a *number* of different people." He grinned.

Maglio pointed to the empty cage again, "Now, make Three Hundred Thirteen get into that cage and be prepared because he will *not* want to go. Then ..."

"What do you mean?"

"You are pretty stupid for a smart kid."

"What do you mean?" I repeated.

"See."

"Hey!"

Maglio shrugged. "Let us just say that the cages hurt them a little."

I looked over to Three Hundred Thirteen and then over to Sixty-One. "A little?"

"Okay, a lot. Now, make him get in that cage and then we will see if the two of us together are strong enough to integrate the numbers."

"And if I refuse?"

Maglio laughed like a possessed doll from one of those really bad horror movies. "I lied about the poison that I gave Kelsie. It was not sleeping poison, Matt. I gave her a fatal dose and she will need the antidote in ..." he paused to look at his watch, "... approximately ten minutes if she is going to live. That poison is working through her system as we speak. Every second you waste is literally killing her cell by precious cell. If you do not hurry, Matt, she is going to have permanent brain damage."

I didn't know if he was lying, but even if he was, I didn't have many options. "Where's the antidote?"

Maglio patted his pocket protector.

I had a quick vision where I suddenly sprinted forward and did some kind of super flying kick and hit Maglio so hard that I knocked him out, the antidote popping out of his pocket and landing neatly in my hand. Then, as Three Hundred Thirteen somehow let all the other numbers out of their cages, I ran back and saved Kelsie.

But that was just a daydream and I was faced with reality. I looked over to Three-Hundred Thirteen. The thought of causing him a lot of pain and completing Maglio's grand plan at the same time almost made me sick. But what could I do?

I'd like to say that I had a surprise up my sleeve, or anywhere for that matter, but I didn't. The truth was, I was fully prepared to force Three Hundred Thirteen into that cage and let Maglio do whatever if it meant saving Kelsie's life. Luckily, I didn't have to worry about it.

"Matt, Kelsie is slowly dying," Maglio's voice cut through my thoughts.

"Actually, she's not!" Kelsie's voice echoed through the cabin. I turned to see her standing at the entrance. She was in a fighting stance holding a spear in her left hand and a small stone knife in her right. A breeze was gently moving her hair around and the lighting gave her skin a golden hue. I have never seen her look so beautiful. And angry. And under the circumstances, I'd have to say beautifully angry.

Maglio was actually stunned, "… but the poison …"

"I didn't drink that tea. I faked everything and waited for you to make a move. I knew you were up to something from the very beginning." She hefted her spear and then slid the knife into her belt.

I suddenly felt like I was in an episode of Scooby Doo and expected Maglio to say something like: *And I would have gotten away with it if it weren't for you meddling kids!*

But instead he completely surprised me. "Well, it looks like it's the Genius Professor versus the Nerd and the Warrior Girl."

And he charged.

Me.

Before I could even react, Maglio had closed to what we would call *melee distance* in my gaming club at school. He kicked my legs out from under me and sent me heavily to the ground. Then his club came crashing down on my ribs and I heard a cracking sound. I suddenly found myself curled up in major pain. All I could manage was a vigorous groaning noise.

"And now just the Warrior Girl." He collected himself and faced Kelsie. In my agony I noticed that he, too, was in a fighting stance. This was not good.

I watched Maglio edge forward with cat-like steps. He kept his body angled so that only part of him was exposed to any attack that Kelsie might spring. He was stronger and more nimble than I thought possible.

Kelsie grasped her spear with both hands. She started forward with small, determined steps.

Maglio edged forward. "This is going to hurt, Warrior Girl. This is going to hurt a lot. Put down the spear and you can go over and help your little wimp boyfriend." Ouch, his comment hurt more than my cracked ribs. Well, almost.

Kelsie stopped and her hands lowered to her sides. She looked as though she might set her spear down. They were close now. I thought she was going to give up. For me. But then, in a flash she brought her spear up and slashed at Maglio with the tip. He ducked the blow, but she brought the other end around with the momentum of the swing and aimed for Maglio's feet. He managed to lift his front foot and the swing went harmlessly underneath. With the momentum, Kelsie's defenses were open to attack.

And attack Maglio did. He swung a vicious downward blow for Kelsie's head that would have probably killed her, except that she rolled to one side and popped up on her feet again. I was surprised at how fast she was and reminded myself that if I got out of this alive, never to mess with her.

The next instant was a blur. Maglio swung wildly sideways, but Kelsie pulled back into a crouch and the blow missed her by inches. Instantly, she sprung forward like an angry cat. With two hands on the spear, she hit Maglio on the side of the head. It was so hard that her spear broke in two and Maglio fell backwards to the ground. I felt the impact of his body landing like dead weight which I hoped he was, dead that is. Or at least very, very hurt. But I was surprised once again that he shook his head and started laughing. Laughing!

The impact had apparently knocked the sense out of him, if there was any to begin with, but it also knocked the club out of his hand. Kelsie stepped over him and put the sharpened spear point to his throat. She was *not* amused. "You want to tell me what's so funny before I spear you?" she growled.

"Actually, I do." Maglio kept laughing, "That was just a diversion." In an instant Kelsie spun into a crouch and seemed ready to fight anything. But she was too late. Fifty-Seven darted from the shadows and went into her, just as it had with Maglio.

Kelsie stopped suddenly and her hands fell to her sides, the spear dropping to the floor. Her eyes went glassy and it looked like she was staring at something far, far away. Her jaw opened as if she were going to scream, but she didn't. She started shaking and then her eyes rolled up so that I could only see the whites.

And then the transformation happened. Her hair dulled and went partially gray. Her flesh rippled and her cheeks sagged. Lines appeared around her eyes and on her forehead, a few at first and then a whole series of them like someone was drawing on her with charcoal or something. Her shoulders slumped forward and her fingers thinned.

Her eyes dropped back down and she was, well, old. She lifted her hands and looked down at them in shock. I was shocked too. I know that age is just a number, at least that's what they say. Fifty-seven isn't that old, but to a teenage kid, it's practically ancient, especially when you see someone you really care about suddenly fast forward to being older than your mom.

The next thing I knew Maglio was up again and had the club in his hand. He hit Kelsie in the stomach and she doubled over. Then he knelt down and shoved her face into the ground where she remained choking for air.

Maglio stood up and circled around me. "How do the ribs feel, Matt? You like numbers, right? Well, if my blow was precise enough, I do believe I just broke your sixth, seventh and eighth ribs. Hurts when you breathe, doesn't it?" I nodded and even that hurt.

"Looks like Maglio wins yet another round."

25

Age Is Just A Number

Maglio continued talking, "You are not the only ones to come through the portal, you know. It all started with my fool colleagues who had no idea what we had discovered. They thought we had simply found an alternate Earth—the geodetic studies and star analysis confirmed it was Earth—and that was amazing enough! Imagine the untapped resources and land! An untainted Earth alone would be worth quintillions." He stopped talking for a moment, caught up in some small daydream.

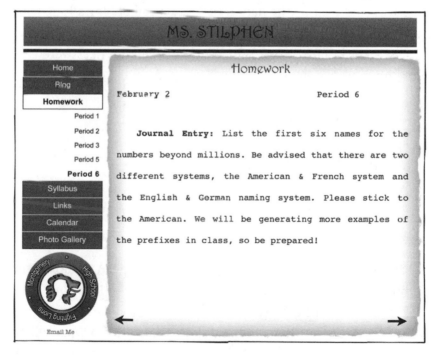

MS. STILPHEN

Home
Blog
Homework
Period 1
Period 2
Period 3
Period 5
Period 6
Syllabus
Links
Calendar
Photo Gallery

Email Me

Homework

February 2   Period 6

Journal Entry: List the first six names for the numbers beyond millions. Be advised that there are two different systems, the American & French system and the English & German naming system. Please stick to the American. We will be generating more examples of the prefixes in class, so be prepared!

The names for numbers in our system follow a consistent pattern. We use commas, for example, every three place values to separate out the digits and make them easier to track. Because we have ten possible digits in each of those three place values, the commas separate groups of $10 \times 10 \times 10$, or 1,000. The number one million is written out as: 1,000,000. That's a 1 followed by six zeros with two commas separating. The word *million* comes from the word for *thousand,* which might seem confusing, except that a million is a thousand thousands, or $1,000 \times 1,000$. Notice the six total zeros. This is also the same as $1,000^2$, or one thousand squared. A billion is a 1 followed by nine zeros (and three commas) and can be thought of as a thousand millions, or $1,000 \times 1,000 \times 1,000$, or one thousand cubed: $1,000^3$. A trillion is then a thousand billions and a quadrillion is one thousand trillions, or a 1 followed by fifteen zeros.

I was confused by the names of these numbers for a long time because I couldn't make sense of them, especially when I learned that million came from the word for thousand. The prefixes were easy enough since:

    bi = two
    tri = three
    quad = four
    quint = five
    sext = six
    sept = seven
    oct = eight

So, here's the system I worked out in sixth grade. A million is the number that results from the *first time* we multiply 1,000 by 1,000. The prefix *bi* means two, and it refers to the *second time* a group of 1,000 is multiplied by a thousand. A trillion is the third time, quadrillion the fourth time and so on. Writing everything out as words and numbers made everything much easier for me:

First: One thousand × 1,000 = one million

Second: One million (one thousand × 1,000) × 1,000 =
one billion

Third: One billion (one thousand × 1,000 × 1,000 × 1,000 =
one trillion

Fourth: One trillion (one thousand × 1,000 × 1,000 × 1,000)
× 1,000 = one quadrillion

Fifth: One quadrillion (one thousand × 1,000 × 1,000 × 1,000
× 1,000) × 1,000 = one quintillion

And so on. By the way, a googol represents the number 1 followed by one hundred zeros, which is a huge number, except that it is dwarfed by the insanely large googolplex, which is a 1 followed by a googol zeros, or $10^{10^{100}}$. This number is so staggeringly, tremendously huge that Carl Sagan supposedly estimated: if the entire known universe were filled with paper full of regularly sized printed zeros, there still wouldn't be enough space *in the universe* to finish writing all the zeros!

I realized Maglio was now mumbling to himself, "… but such a thing would require an army and …" he shook his head. "Ah, Maglio digresses once again. In the beginning, we could not decide what to do. They had such pitiful ideas, such pathetic ambition," he frowned and shook his head. "And such allegiance to an evil empire that exploited their every thought! It was too easy to get rid of them."

Kelsie moaned. I looked over and she was beginning to stir a little. Maglio walked over and stepped on her, pinning her to the ground. "Others came through the portal as well and I quickly realized that there were changes that took place here based on what people came through. Sometimes the weather would change, sometimes the grass or the clouds. In your case, it was the snowflakes that alerted me."

He was about to reveal more of his scheming plans, just like the villains always seem to do in books and movies, but I finally had my idea. "How old are you, Maglio?"

He turned and looked at me as if I was up to something, which I was. He stepped off of Kelsie and walked a few steps toward me. He started to respond, "I'm thirty …"

In that moment I willed Three Hundred Thirteen into Maglio. I thought about Kelsie and all the time I still wanted to spend with her. I thought about my family and friends. I thought about our world and Maglio stealing everything. I thought about this world and the imprisoned numbers. Then I thought again about Kelsie.

Three Hundred Thirteen hadn't been able to communicate with me yet, but I willed him to attack Maglio with every bit of determination I had. Somehow he understood. He ghosted into Maglio and instantly, there was a battle of wills. I was trying to force Three Hundred Thirteen completely into his body and Maglio was trying to reject him. My mind was suddenly flooded with numbers and equations.

$300 + 10 + 3 =$

$626 \div 2 =$

The square root of 97,969 =

Donald Duck's license plate …

Maglio's eyes went glassy. His mouth quivered, "I'm thre…"

$CCCXIII =$

$10011100\mathit{1}_2 =$

$25^2 + 312^2 = c^2$

$471_8 =$

Maglio started shaking. "I'm three hun…"

$(4 \times 78) + 1 =$

The twin prime of 311 =

$1^1 + 2^2 + 3^3 + 4^4 + 5^2 =$

$12^2 + 13^2 =$

His eyes rolled back. "I'm … I'm three hundred thirteen …"

Matt,

The second Career Day is coming up and, because you turned in your form late, I couldn't get you into any of your first choices. The Electrical Engineering workshop was booked on the first day! I managed to squeeze you into your eighth choice, which I think is pretty good under the circumstances. On Tuesday at 11:00 report to the Embalming Workshop in Room #12C. Have a great time!

Mr. Nguyen
Senior Advisor

Then the transformation occurred. At first it looked similar to the changes that he had undergone before, but they did not stop at wrinkles and gray hair. His mouth and lips swelled. Red and green streaks marbled his skin as if he were covered with tattoos of spider webs. Then his entire body bloated until he was nearly twice his normal size. His clothes burst and shredded, with the exception of his nano-shirt, which expanded with him. Brown and black blisters erupted on his skin leaving holes in his body. Drops of blood and other fluids fell to the ground and splattered into dry powder.

A cloud of greenish brown gas burst out of his chest and the shock sent bits of his hair floating to the ground. Then he began shrinking. His eyes sank into their sockets and his face collapsed. His marbled skin dried almost instantly, then flecked off and scattered like bits of gray paper. His flesh shriveled into white waxy clusters and then even those clumps disappeared until all that remained was his skeletal structure. Ligaments held his bones together briefly, but I watched as even those shriveled and disappeared. His bones stayed in place for an instant and then his skeleton collapsed into a pile of what is usually something like 206 human bones.

I saw Three Hundred Thirteen floating in the place where Maglio had just been a moment before and then my vision faded and I blacked out.

26

Return

When I regained consciousness, my first thought after being completely amazed at Maglio's accelerated and extremely gross death, was about Kelsie. She had taken quite a blow to the stomach and I hoped that she didn't have any broken ribs like I did. I rolled over painfully and then propped myself up. The pain was intense, and I coughed up some blood, but I managed to crawl over to her. She had returned to her normal age and I was relieved to find that she was breathing regularly. I checked her pulse and she seemed fine. I grasped her hand between both of mine and gently tried to wake her, but she was totally unconscious.

I scanned the room for Fifty-Seven and finally saw him hidden behind the teleportal. I wasn't sure what he was doing there, but figured with Maglio gone he didn't know what to do. Perhaps the other numbers thought of him as a traitor and he was hiding or something. Three Hundred Thirteen was flitting around the cubic generator.

I hobbled to the center of the room and searched the cube for a power switch. I didn't find one, so I just yanked out all of the wires, including a few that led up to the roof. The humming noise faded and the holes in the cube iterated closed. In a few seconds the whole thing was a solid mass of metal. I opened all of the cages and then watched as most of the numbers ghosted away out of the cabin. Sixty-One floated over to Kelsie and if his eyes were human, I would have said that he looked worried. Three Hundred Thirteen stayed by my side.

I hobbled into the kitchen area and saw the can of pickled herring again. I passed the pile of garbage, and nearly puked, which wouldn't have been so bad except that it sent sharp pains shooting through my ribs and I coughed up more blood. I walked into the bunk bed room and quickly scanned around. There were eight bunks total, but the five duffle bags and seven trunks were piled on only two of them. There were clothes and papers and food wrappers spread all around, making my room back home seem clean *and* tidy. I grabbed a rolled-up sleeping bag and the cleanest pillow I could find.

I limped back toward Kelsie and couldn't help grabbing the can of pickled herring. I unfurled the sleeping bag and then unzipped it to form a blanket, and then I placed it over her. I gently lifted her head and placed the pillow underneath.

Now I had a most important task: to find a can opener.

After cleaning and tidying up the place, I threw a tarp over Maglio's remains. I stared down at the pile and it suddenly dawned on me just what had happened: I had been responsible for killing someone! I went through the events in my head and convinced myself that it was all self-defense. That made me innocent, right?

I didn't really want to think about it anymore and I was still worried about Kelsie, so I grabbed the cards off the kitchen table and played Solitaire on the floor next to her for a while. I tried to keep my mind distracted from what I had done. And I tried the pickled herring for the first—and last—time of my life.

A deck of cards brings to mind all kinds of mathematics, from simple arithmetical tricks to the probabilities of counting cards in multiple decks in high stakes poker. I never got into gambling, maybe because so many of my friends and family keep annoying me about how good I could be.

Although I love a good card game or mathematical card trick, two main features of a deck have always caught me. A standard deck of cards has four suits of thirteen cards, totaling fifty-two, plus two jokers. There are fifty-two weeks in a year, and I have always wondered if a particular card goes with a certain week, sort of like a weird astrology or something. A specific card could mark every Sunday, so that we could refer to weeks as "the week of the eight of clubs" or "the week of the king of diamonds."

But it is the number thirteen that fascinates me the most, and it is partly why my favorite number is 313. A lot of people think that the number thirteen is unlucky, but I'm not sure why. The superstition is so widespread that there is even a special word for fearing the number: *triskaidekaphobia*. There are some airlines that don't have a 13th row of seats, just as there are places where the street numbers skip thirteen. I know that some buildings skip the thirteenth floor (just look in the elevators), which is weird since there still is a thirteenth floor, even though they are not numbered that way. There is the superstition of Friday the 13th, of course, and card 13 of the tarot is the death card— though I've been told that it doesn't always refer to actually dying.

The fear of the number might trace back to a Norse myth about twelve gods having a dinner party. The trickster god Loki came and shot one of the other gods with an arrow and plunged the world into darkness. Others believe that on a Friday the 13th the last 13 Knights Templar were killed. Supposedly, there were thirteen witches hanged in the witch trials in Salem, Massachusetts. Others feel that the superstition comes from The Last Supper, where Jesus announced that he would be betrayed and that his disciples would abandon him.

One example why thirteen might be seen as unlucky is because of the Apollo 13 space mission, which launched at 13:13 CST. The mission almost met with complete disaster when an oxygen tank blew on April 13, 1970, but through incredible resourcefulness, the crew survived. In this case, thirteen seems more like a number for perseverance and ingenuity, not bad luck.

Thirteen seems like a positive number in the history of the United States. There are thirteen stripes on the U.S. flag representing the original thirteen colonies. On the one dollar bill there are 13 bars on the shield, 13 leaves on the olive branch, 13 arrows, 13 stars, 13 steps on the pyramid and even 13 letters in both ANNUIT COEPTIS and E PLURIBUS UNUM, and I'm sure many more thirteens that I have yet to find. The 13th Amendment abolished slavery.

At Colgate University, the number is actually considered lucky. The school was founded in 1819, as the story goes, by "13 men with 13 dollars, and 13 prayers." The school address is 13 Oak Drive and the zip code there for Hamilton, NY is 13346, which can be thought of as 13 and 3+4+6, which is another 13. One of the a cappella groups there is known as the Colgate 13.

Overall, I have found more examples of 13 being positive rather than unlucky.

I was about to win my thirteenth game of Solitaire, when Kelsie awoke. "Hi, handsome!" she whispered, "I take it we won."

"How are you feeling?"

She just nodded.

"Yeah, we won," I said.

She sat up, "Uh, my head aches." She tried to run her fingers through her hair, but there was too much caked blood. She moved around, testing her joints and stuff. "I don't think anything is broken. How about you?"

"I think I have three cracked ribs," I said grimacing. "It hurts when I breathe, and I keep coughing up blood."

Kelsie looked around at all of the empty cages. She smiled when she saw Sixty-One and Three Hundred Thirteen. "What happened to Maglio?"

I motioned to the tarp with my head and then told her everything.

"So, let me get this straight: Maglio and his colleagues made the teleportal thing, they came through and set up this observation cabin, or whatever this is. Then he figured out the whole numbers thing and killed the other scientists—and other people as well. He set up all this number manipulating machinery—and then basically we arrived?"

"Pretty much."

"And he was forcing you to help him become a zillionaire or whatever?"

I almost answered that *zillionaire* wasn't a real word, but instead replied, "Well, crazy rich, yeah."

She shifted and then I helped her to her feet. She flexed and twisted slowly. "Ow, that hurts."

"What does?" I asked, concerned that maybe she really did have some broken bones.

"Everything. But I'll be better when I can clean up." She tilted her head, batted her eyelashes, and ran her hand down her blood-crusted cheek, "Unless you like my new make-up."

"Eyew, Kelsie, that's gross."

She laughed. "Well, I've had worse," she declared. "This is like an average day with my brothers."

I cringed because I could remember the time when she nearly broke her arm when we were playing tackle football, sprained her knee when we were skiing, broke two fingers when her brother Mark pushed her off her bike and a whole series of other events.

She hobbled toward the pantry area. "Is there anything good to eat? I'm starving."

I chuckled to myself and then said, "You should try the pickled herring. It's, uh, *really* good."

"Really?" she said enthusiastically. "I've never tried it. What's it like?"

"Well, uh, I think you should just have some." I fished a can of the herring out of one of the crates and grabbed the can opener. "Here," I said, handing her a spoon.

I was excited to ambush her taste buds with the grotesque stuff, but she shoveled spoonful after spoonful into her mouth. "Mmm, you're right, Matt, that was good." She winked at me and set the can down. "Ah, much better."

I was speechless because I had nearly thrown up at the mere smell and spit out the spoonful I had tried earlier. I ended up eating a couple of cans of beets.

We spent the next few days recovering. We thought about going back, but considering the condition we were both in, neither one of us wanted to go through the teleportal. The first trip had been painful enough, but with cracked ribs and assorted injuries, I know I wasn't ready. We played a bunch of games that we found in one of the trunks, but it seemed like no matter what game we played, I somehow won 313 to 61. Kelsie was good-natured about it, but she did accuse me of cheating on several occasions.

After lengthy discussions, Kelsie finally convinced me that I was one hundred percent innocent and I put the matter to rest. She even tried to convince me that I was a hero, but I couldn't go that far. I thought she was the real heroine, considering that she had suspected Maglio all along and had stood up to him, whereas I had mostly just reacted. In any event, like a lot of science fiction and fantasy villains, Maglio had been a murderer bent on world domination and the two of us had stopped him.

We talked about home a lot and what we thought everyone was doing. We couldn't help thinking about how much trouble we might be in, especially considering we didn't know how much time had elapsed. Maglio had mentioned going through the teleportal a few times, so there probably wasn't a huge difference—at least I hoped. I could imagine returning to find that twenty or thirty years had passed and our friends and family were all old. Or perhaps worse, that we returned to ten years earlier and all of our friends were annoying eight year-olds.

We took some time and tried to teach both Sixty-One and Three Hundred Thirteen to communicate, but we were mostly unsuccessful until Fifty-Seven came out of hiding and helped. We retrieved Kelsie's iPhone and used the QR encoder to send messages back to them. At first, we could only communicate with Fifty-Seven and he somehow translated for the other two. After a few days, though, Sixty-One and Three Hundred Thirteen got the hang of it.

They were more intelligent than I thought, but nothing like talking to other people—more like dolphins or something. They could only communicate simple concepts using a few words. I realized after a while that when Fifty-Seven had "spoken" to us several days ago, he had mainly been mimicking what Maglio had taught him to say. What I grew to appreciate was that they each had their own personalities. Fifty-Seven was timid and soft-spoken, while Sixty-One was energetic and "talkative," and Three Hundred Thirteen was inquisitive and eager to learn.

When we finally felt ready, we gathered all of our stuff and stood by the teleportal. We said our goodbyes to the numbers and watched as they coded back to us: "Goodbye." I wasn't sure if they really understood what was happening, but there wasn't much I could do. I threw the switch and the torus hummed to life.

Kelsie grabbed my arm and looked back at the cages, "It's tempting, isn't it, Matt?"

"You mean the lotteries and everything?"

"Yeah."

"I don't know, Kelsie. They say money can't make you happy."

"Yeah, but I wouldn't mind trying to disprove that." She looked over to Three Hundred Thirteen and Sixty-One, "Maybe they could somehow just get us a little money for college on scratch tickets or something…"

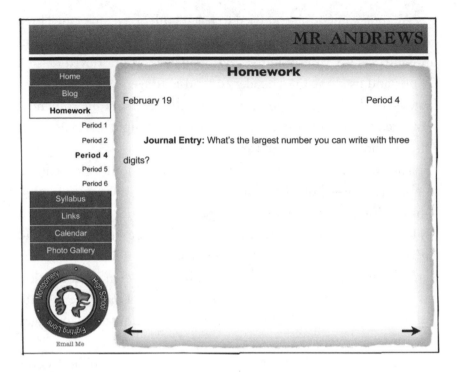

Homework

February 19   Period 4

Journal Entry: What's the largest number you can write with three digits?

A lot of my classmates answered 999, or nine hundred ninety-nine. But there are much larger numbers that can be generated with three digits. Here's where exponents become powerful. 99^9, or ninety-nine to the ninth power is much larger, and nine raised to the ninety-ninth power, 9^{99}, for example, is larger still. But the sequence of exponents can keep going by raising the second nine to another power, so that the final answer is tremendously huge. So, the largest number that can be written with three digits is nine to the power of nine raised to the ninth power, or 9^{9^9}. Although such a number is already incredibly large, if other symbols were allowed, there could even be $9!^{9!^{9!}}$.

Kelsie's idea of combining 61 and 313 together into $61,313 would certainly be nice, but it wouldn't even pay for us to go to college these days. Besides, far larger numbers could be generated. But I saw the way the numbers were reacting inside the cages and the way they darted out when I let them free. "No way, Kelsie, the machines hurt them too much. I could never be responsible for that."

She sighed, "I know, Matt. I was just imagining."

I smiled, "But I do have some good news. They agreed to help a little."

We held hands as we stepped through the teleportal.

Epilogue

The trip back through the teleportal was even more painful than the first time, mainly because we were already injured. Pain burned through my chest and rose up into my head like an electrical storm. Weird geometrical fireworks erupted behind my eyes and then my vision darkened. I felt my body collapse and then I blacked out.

I expected to wake up in the lab with Kelsie standing above me, but I was unpleasantly surprised to wake up in a jail cell. My groggy head cleared in a fraction of a second as I realized where I was. "Ah, Mr. Forsythe, I'm glad you can finally join us," a strange voice said. Some guy in a suit opened the cell and led me to one of those barren questioning rooms with one-way glass and a table bolted to the floor, like in all the crime movies. I wondered if my parents were behind the glass. A woman at the table stood up as I entered, "Hello, Matthew, I'm your court-appointed attorney."

"Where's Kelsie?" I asked.

"You have other things to worry about, Matthew," she stated.

I would like to say that things got better from there. I wish the space-time difference had been barely noticeable, like perhaps we had actually only been gone a few seconds from our world. But the truth was that almost nine days had passed. The police had gotten involved at that point, since our parents had reported us as missing, and considering the building our school was now in, the FBI got involved as well.

Several men in black suits crowded into the room. One of them stepped forward, "Mister Forsythe, you are in a great deal of trouble!"

I'm not sure what I expected, but "Thanks for saving the world, Matt" or "Thanks for dispatching a master criminal" were much more in line with what I was thinking. Instead, I was asked a zillion questions (a word that suddenly seemed suitable to me) by so many FBI agents and police officers that I felt like my head was spinning like that possessed girl in *The Exorcist.*

They did the whole good cop / bad cop thing and tried to mess with my head by insisting that Kelsie and the others were ratting on me for everything. I just stuck to my story, trusting, or hoping, that truth would win out—even with potential murder charges.

I explained everything, especially the fact that the numbers were alive and Maglio had been manipulating them to take over the world. The police thought I was crazy, which I guess is understandable, but the men in black suits were remarkably interested. The FBI quickly took over the entire investigation, and I was shuttled from one black van to another, one building to another. I explained over and over again to different groups of people what had happened, starting with the sink in the bathroom.

The FBI confiscated everything, including Kelsie's iPhone, the *upturn* table and even the classic rock albums, which nearly made Jamie and Thomas cry. They shut down our school building and we all celebrated at first thinking that school would be canceled for the rest of the year. But somehow the Superintendent found unstructured space in an old ketchup factory. Construction started immediately, but it was a while before the space was ready. According to state law, we still had to make up the days, so the school calendar had to be bumped forward all the way into late August and we even had to attend classes on Saturdays. We had to spend the last sixty-one days of school in makeshift classrooms with no air conditioning in the middle of summer—in an old ketchup factory!

Our school administered consequences, which bothered me a lot, since *rewards* seemed like a better word for what had occurred. Ari, Jamie, John and Thomas only got three weeks of detention for skipping classes, but Kelsie and I had a lot to answer for. We got six weeks of detention and had to be escorted everywhere when we were on school property. During study hall, we had to do community service scanning books into the library database. *Boring* is an understatement compared to the adventure we had just been on.

Our grades suffered a bit, since we missed a few tests and quizzes and the school administrators refused to allow us to make them up. Kelsie's average dropped six tenths of a point, and she lost her status as salutatorian to Jessica Doolittle. "Geez, at least her name could have been like *Crafty* or something," she complained. My average dropped a little as well, but nothing significant resulted.

All the official charges were dropped and the FBI declared publicly that the six of us were basically nothing more than school hooligans, or *schooligans* as Ari dubbed us. We all had to sign "special papers," but I can't say anything more about them. The cover-up story became that Kelsie and I had tried to elope by first hiding in the school. We supposedly planned on flying to Las Vegas to get married.

I saw my parents eventually, but they were a terrifying mix of angry and delighted to see me. After hugging me profusely, they informed me I was grounded "until doomsday" and that I couldn't see Kelsie ever again. When I finally convinced them that Kelsie and I had *only* saved the world (including their retirement plans and my college fund), they went easier on me. Kelsie was grounded for "forever and a day," but once things settled down and her parents cooled off, she was free again.

The police also assigned us community service, since they had spent time and resources looking for the two of us. I thought maybe we would have to teach young kids chess or something, but apparently someone in the police department thought it was fitting for us to clean bathrooms, since everything had started there. So, Kelsie and I—savers of the entire world—had to scrub public toilets twice a day for six weeks at Quentin Park, or Conse-Quentin Park, as we got to calling it.

Kelsie and I finished at the park one day, and we were sitting on a bench waiting for our rides when she said, "So Matt, I understand that Fifty-Seven helped us get a new school building, which, I guess, was a good thing. And Sixty-One sort of helped us finish high school four days before we have to go off to college and all. But, you know, I thought they were *really* going to *help* us."

"You mean money or something?"

"Yeah."

"Finishing school is important," I pointed out.

"Obviously. I just thought maybe we could have been rewarded a little for what we did."

I leaned over and kissed her. "Yes, of course. I just wish we had spent more *time* there together." She smiled at me.

"Besides, Kelsie, Three Hundred Thirteen helped us out enough."

Beyond Infinity
A MatheMATTical Adventure

SENIOR SYNTHESIS PROJECT
MATTHEW FORSYTHE
PRIMARY ADVISOR: MR. NGUYEN
~~JUNE 14, 2012~~
AUGUST 15, 2012

Primary Advisor: _____Mr. Nguyen_____

Matt,

I certainly enjoyed reading your project. It was full of little gems and insights about math that were woven into a fun narrative. I particularly like the part about cardinal, ordinal and the combined numbers. That was a lot to think about! I also like the idea of the numbers being alive. It gave me a whole new perspective on thinking about "the secret lives of numbers."

In terms of the synthesis of your mathematics career here, this manuscript met all of the requirements of the math department Interestingly, you also appeared to meet most or all of the English department requirements as well. Are you sure you don't want to try creative writing at college? At least give it a thought!

I'm sure glad all of those teachers were made up! I certainly wouldn't want to be some of them. By the way, I shared the concept of "the window" with the superintendent, and she has assured me that the idea will be brought up in future faculty meetings. I'm glad you made the right choice and didn't sell it to other students.

Final Critical Comments:

My main criticism is that you didn't include more from this year of advanced calculus. I know you must have been busy this semester thinking about college, finishing your work, running those clubs and everything, but more of the most recent math would have provided a broader synthesis of what you really learned.

GRADE: _____A-_____

Secondary Advisor: _____Mr. Watson_____

Matt,

This final version was clearly the best. This independent study met all of the standards for the creative writing requirements for the half-credit of English. I enjoyed reading the text, but I have to admit the journal entries and puzzles frustrated me. I'm glad your explanations were thorough enough for me to understand at least some of the math.

What excerpt are you going to read at the fair? Remember, you have a twenty-minute slot to fill, so make sure you practice the timing. Also, you should be thinking about what section to read at the actual graduation ceremony. This should be no more than ten minutes.

Final Critical Comments:

There are a number of changes that I wish you had made, based on the last meeting Mr. Nguyen and I had with you. In particular, he wanted to see more from this year in calculus in order to give a more complete picture of your final year. As you know, I wanted to see you edit more of the fragment sentences out of the manuscript. I know you keep insisting that fragments are part of your voice, but I felt as though there were too many. I'm still not sure about some the puzzles. Remember, you have to consider your whole audience.

GRADE: _____A-_____

Peer Advisor: Jeremy Morris

Yo Matt,

Awesome project, man! I can't believe there's a secret door in the bathroom! I was really impressed with the amount of work you put into this. It makes my robot project look a little ... light ... get it? Anyway, I thought the ratio of the amount of math to the narration was just right in the final draft. We had spoken on a number of occasions on the earlier drafts and I'm glad you took my advice and edited some of the math out. I know Mr. Ngyuen wanted more calculus in there, but honestly, for the average reader, it was too much—especially the part with calculating volumes of solids around the X-axis. Way too much!

What are you and Kelsie going to do now that you're going to different schools? :o

Final Critical Comments:

I don't have any critical feedback at this point, since we already went over everything in earlier drafts. As I said before, I disagree with Mr. Nguyen about including all the calculus. And I disagree with Ms. Watson about the fragment sentences. They. Didn't. Bother. Me. At All. You made all the changes we talked about, so I have nothing else to add.

GRADE RECOMMENDED: __99.9__ (Hey, no one's perfect)

P.S. — Did you and Kelsie really get $97,969 each in scholarships?

Please Note:

1. No squirrels, rabbits, acorns or even mad scientists were hurt during the composition of this synthesis project.

2. The number two is not actually a cobra or a swan.

3. No real-life teachers or students were mocked or ridiculed.

4. Square numbers do not square dance.

5. The number 13 did not attempt to bribe me.

6. The number six is not trying to sue me for defamation.

7. I am not under duress when I claim that school lunches are actually delicious. De-licious!

8. Skipping school to seek cool adventures is wrong. Very wrong.

9. 9•6—25•15•21—1•18•5—18•5•1•4•9•14•7—20•8•9•19: 7•5•20— 1—12•9•6•5

Special Thanks To:

Jim Andrews
For teaching me so much about mathematics.

Aspen Academy
For being so supportive of being entrepreneurial.

Ross Blackburn
Evan Breitkreuz
For reading an early draft and keeping me focused.

Judith Briles
For all her wisdom about books and publishing.

Judy Fischer
For always supporting my endeavors.

Santana Garcia
For reading an early draft and keeping me focused.

The Greenes
For all their support over the years.

Mary Kelley
For making suggestions on my early drafts.

Kickstarter
For helping launch great projects.

John Maling
For editing and assisting.

Susan Midlarsky
For introducing me to National Novel Writing Month.

Brian and Sharon Parker
For the close reading and comments on an early draft.

Rachel Perry
For making suggestions on my early drafts.

Gary Rider
For encouraging me to become a writer.

Jeremy Rozen
For his feedback early on in the process.

Louise Stilphen
For her constant support over many years.

Becky Southwick
For being one of my earliest readers.

Sparhawk School
For being such an amazing place for teaching and learning.

Conor Sullivan
For making suggestions on my early drafts.

Nan Washburn
For encouraging me to become a writer and teacher.

Nick Zelinger
For designing the cover and the interior.

All of my current, former and future students …
For making teaching and learning so enjoyable!

About the Author

Charles Ames Fischer is a teacher by day and an author by night. He enjoys spending time outdoors, reading good books, watching hockey games and contemplating the meaning of life. If pushed, he might admit to spending way too much time playing fantasy games. He taught for many years at Sparhawk School in Amesbury, MA and currently teaches at Aspen Academy in Greenwood Village, CO. Throughout the year he conducts a variety of fun-filled workshops for students and professional development seminars for educators. This is his first novel (with several on the way).

Learn more at: *www.CharlesAmesFischer.com*

Visit the book's website *www.BeyondInfinityBook.com* to learn more about math and numbers, and to get involved in a unique community. Read some bonus chapters that didn't make the final cuts, add your own number fact examples and vote on the best ones, find out the popularity of people's favorite numbers, participate in fun challenges, and much more!

Charles Fischer, By the Numbers (2013):
    Years as a writer: 32
    Number of novels planned: 4
    Years as an educator: 18
    Favorite number, rational: 64
    Favorite number, irrational: π
    Countries visited: 12
    States visited: 47
    Number of states lived in: 5
    Favorite Harry Potter, book: 3
    Favorite Harry Potter, movie: 6
    Favorite Star Wars movie: 5

Favorite Lord of the Rings: 2
Number of cars owned in lifetime: 4
Times ridden in 3-wheeled truck: 1
Times in 3-wheeled truck accident: 1
Egyptian pyramids visited: 9
Number of pyramids where friend disappeared: 1
Amazing teaching & learning moments: Beyond Infinity

A MatheMATTical Adventure A MatheMATTical Adventure
A MatheMATTical Adventure
A MatheMATTical Adventure
A MatheMATTical Adventure